Environmental Pollution
and Control

Third Edition

"If seven maids with seven mops
  Swept it for a half a year,
Do you suppose," the Walrus said,
"That they could get it clear?"
"I doubt it," said the Carpenter,
  And shed a bitter tear.
                              —Lewis Carroll

Environmental Pollution and Control

Third Edition

by
P. Aarne Vesilind and J. Jeffrey Peirce
Department of Civil and Environmental Engineering
Duke University
Durham, North Carolina
and
Ruth F. Weiner
Huntley College of Environmental Studies
Western Washington University
Bellingham, Washington

Butterworth–Heinemann
Boston • London
Sydney • Toronto • Wellington

Library of Congress Cataloging-in-Publication Data

Vesilind, P. Aarne.
    Environmental pollution and control / by P. Aarne Vesilind and
J Jeffrey Peirce and Ruth F. Weiner.—3rd.
      p. cm.
    Includes bibliographical references.
    ISBN 0–409–90272–1
    1. Environmental engineering. I. Peirce, J. Jeffrey.
II. Weiner, Ruth F. III. Title.
TD145.V43 1990
628.5—dc20   90–1657
                                          CIP

British Library Cataloguing in Publication Data

Vesilind, P. Aarne
    Environmental pollution control.–3rd ed.
    1. Environment. Pollution. Control measures
    I. Title II. Peirce, J. Jeffrey III. Weiner, Ruth F.
    628.5

    ISBN 0–409–90272–1

Butterworth–Heinemann
80 Montvale Avenue
Stoneham, MA 02180

10 9 8 7 6 5 4 3 2 1

Printed in the United States of America

to

Pam, Steve, and Laurie B.

and

Shayn and Leyf

and

Lisa, Annie, Sarah, and Rachel

Contents

Preface

The objective of this book is to package the more important aspects of environmental engineering science and technology in an organized manner and present this mainly technical material to a nonengineering audience. This book originally began as a set of class notes for a course offered at Duke University by the Department of Civil and Environmental Engineering. The course is designed for nonengineering students and has been a popular elective. The third edition owes some of its development to an undergraduate course offered since 1976 in the Huxley College of Environmental Studies at Western Washington University. Although the courses have no prerequisites, we assume that the student has a knowledge of high school or college freshman chemistry for nonmajors. A knowledge of college algebra is required, but calculus is not used.

We do not intend for this book to be scientifically and technically complete. In fact, many complex environmental problems have been simplified to the threshold of pain for many engineers and scientists. Our objective, however, is not to impress nontechnical students with the rigors and complexities of pollution control technology but rather to make some of the language and ideas of environmental engineering more understandable.

P. Aarne Vesilind
J. Jeffrey Peirce
Ruth F. Weiner

Environmental Pollution and Control

Third Edition

Chapter 1

Environmental Pollution

The pictures of the earth from the manned space missions clearly illustrated that not only is the earth round but that it is indeed a finite and lonely little blob. Somehow the sight of this forlorn spaceship, floating friendless in the blackness of space, brought home the fact that the earth and its natural resources are indeed all we have and that we must start worrying about the future of this earth.

The present generation is the first to include among its list of concerns the very survival of the planet, a question that would have been meaningless not too long ago. Before Copernicus not only was the world flat it also was the center of the entire universe. These past few centuries have witnessed the dramatic change from a human population that thought of itself powerless to change the earth and anything on its face to a generation that can literally blow the earth up if it so chooses.

"Blowing up" can take many forms—from the obvious nuclear holocaust to the less dramatic but equally effective method of increasing the ambient temperature through the greenhouse effect and destroying the delicate balance of life. Put simply, the slow, insidious process of destroying our own home by contaminating and killing the earth's ability to support life is what we call *environmental pollution*.

ROOTS OF OUR CONCERN WITH ENVIRONMENTAL POLLUTION

There are two basic reasons for our concern with environmental pollution: (1) human health and welfare and (2) our care about the remainder of nature. The latter concern is not only for the benefits we might derive from nature but because all of nature has its own right to exist. In this

1

chapter we devote most of the discussion to the first concern, human health and well-being, and turn to the latter reason for environmental concern, care about the remainder of nature, in the very last chapter of this book.

Human health and well-being can be impacted by environmental pollution in two distinctly different ways: (1) on a personal level of detrimental health effects due to contaminated water, air, or food or (2) on a global level as the slow yet progressive deterioration of our habitat, resulting in the eventual destruction of the human species and perhaps all life. Such a concern is not very personal, since the possibility of that occurring in our lifetime is about as probable as the sky falling. (Yet we still have many "Chicken Littles" running around trying to set dates when the world will come to an end.)[1] Our personal concern with environmental pollution revolves around immediate health problems, and there are plenty of them to worry about from chlorinated pesticides through heavy metals and asbestos to contaminated indoor air.

The idea that contamination of the environment is detrimental is not new. Although most ancient civilizations had strict prohibitions against various forms of pollution, it wasn't until the mid-nineteenth century that the idea of taking care of our own environment became a concern to people. Contaminated water and air were recognized as being associated with some diseases, and efforts were made to clean up the cities. For example, a resolution in 1869 by the Massachusetts Board of Health proclaimed:

> We believe all citizens have an inherent right to the enjoyment of pure and uncontaminated air and water and soil; that this right should be regarded as belonging to the whole community; and that no one should be allowed to trespass upon it by his carelessness or his avarice or even his ignorance.

This was a worthy resolution and would serve well as a code of conduct for civilization today. The fact that environmental pollution can make people sick and even prematurely kill people through various forms of environmentally mitigated diseases is a major driving force for environmental pollution control.

On a different level, we deplore the destruction of our "sacred places" in nature because of the detrimental effect this has on our mental health. Many of us enjoy the outdoors, and the wilder and more alone the encounter, the more we like it. We would be very much the poorer if nature were not out there for us to be a part of and to enjoy. But this defense of nature

---

[1] See, for example, Paul Erlich's *"Eco-catastrophe"* (San Francisco: City Lights Books, 1969), which begins with the prediction: "The end of the oceans came late in the summer of 1979. . . ."

is *anthropocentric,* i.e. for the good of the people, not for the good of the rest of nature. This is not to say that such a desire is inappropriate, selfish, or wrong. It is a legitimate desire just as freedom from infectious diseases is a human desire. It must be recognized, however, for what it is, a benefit to humans.

The concern with human health is a tricky problem since it involves the concept of the analysis of *risk.* Everything we do in life involves a risk (even staying in bed—you might get hit by lightning or the building might burn down around you), and we manage risk by reducing it whenever possible and desirable. Before we discuss the concern of risks due to the contamination of our environment, it is necessary to understand the principles of risk assessment.

RISK ANALYSIS

A substance is considered a pollutant because it is perceived to have an adverse effect on the environment and, either directly or indirectly, an adverse effect on human health. It is sometimes difficult to determine, however, if there is an effect or if the effects have been deleterious or detrimental.

For example, we are now quite certain that cigarette smoke is unhealthy. We have specifically identified inhaled cigarette smoke as contributing significantly to lung cancer, chronic obstructive pulmonary disease, and heart disease. Notice that we do not say that cigarette smoking *causes* these health problems because we have not identified the causes—the etiology—of any of them at least in the sense that we have identified the poliomyelitis virus as the cause of polio. How then has cigarette smoking been identified as a contributing factor if it cannot be identified as the cause? Cigarette smoking can serve as a good example of how health effects of pollutants are determined although cigarette smoke is not regulated the way other pollutants are regulated.

During the twentieth century with the widespread use of vaccination and antibiotics, infectious diseases have ceased to be a primary cause of death. The life span of people in the developed countries of the world has lengthened considerably, and heart disease and cancer have become leading causes of death, and it was observed during the early 1960s that lifelong heavy cigarette smokers often died from lung cancer.

But how do we determine what people die of, and how is the death related to a habit like heavy smoking? The cause of death listed on death certificates, which are the source of much epidemiological information, does not specify cigarette smoking as the cause of death. In order to relate death from lung cancer to smoking, one must show that significantly more lung cancer deaths occur in smokers than in nonsmokers. Such a showing is called a *standard mortality ratio* (SMR) and is defined as:

$$\text{SMR} = \frac{\text{observed deaths}}{\text{expected deaths}}$$

The number of "expected deaths" in the above equation is the number of deaths from the particular disease (lung cancer in this case) that seem to happen without any identifiable cause. In the general population of smokers and nonsmokers, there are a certain number of lung cancer deaths. Even in nonsmokers, there are a certain, albeit small, number of lung cancer deaths. In this instance, then, the SMR can be defined as

$$\text{SMR} = \frac{D_s}{D_{ns}}$$

where D_s = lung cancer deaths in a given population of smokers

D_{ns} = lung cancer deaths in a nonsmoking population of the same size

In this instance, the SMR is 11/1. Since the SMR is significantly greater than 1.0, we can say that cigarette smoke contributes significantly to lung cancer, or as it is usually phrased, smoking cigarettes significantly increases the risk of death from lung cancer. To be precise, the risk of death from lung cancer is eleven times as high for a heavy smoker as for a nonsmoker. Determination of the SMR tells something about the *epidemiology* of smoking, but not about its *etiology*.

There are three characteristics of epidemiological reasoning in this example that are important:

- Not everyone who smokes heavily will die of lung cancer.
- Some nonsmokers die of lung cancer.
- Therefore one cannot unequivocally relate any given individual lung cancer death to cigarette smoking.

Risk also can be expressed in other ways. One commonly used statistic in risk analysis is the number of *deaths from a given cause per 100,000 population*. For example, in the United States there are 350,000 deaths each year from lung cancer and heart disease that are attributable to smoking. The United States has a population of 240 million. The risk of death associated with the effects of cigarette smoking can thus be expressed as deaths per 100,000 population or

$$\frac{350,000}{240 \times 10^6} = \frac{146}{100,000}$$

Table 1-1. Approximate Adult Deaths Annually per 100,000 Population

Cause of Death	Deaths/10^5 Population
Cardiovascular disease	408
Cancer	193
Chronic obstructive pulmonary disease	31
Motor vehicle accidents	18.6
Alcoholism	11.4
Other causes	208
All causes	870

In other words, a heavy smoker in the United States has an annual probability (or risk) of 146 in 100,000 of dying of lung cancer or heart disease.

Table 1-1 presents some typical mortality statistics for the United States.

A third way to present risk statistics is as *deaths from a given cause per 1000 deaths*. For example, there are approximately 2.2×10^6 deaths annually in the United States. Three hundred and fifty thousand of these, or 160 per 1000 deaths, are related to heavy smoking.

In summary, risk of death due to some environmental cause can be expressed in three ways:

1. Risk of death from being exposed to a given environmental pollutant (SMR)

$$= \frac{\text{number of deaths from a specific cause in a given population exposed to an environmental pollutant}}{\text{number of deaths from the same cause in a similar sized population not exposed to that pollutant}}$$

2. Risk of death from a given cause

$$= \frac{\text{number of deaths associated with the cause in a given time}}{\text{total population, all of whom will die due to } some \text{ cause}}$$

3. Risk of death from a given cause

$$= \frac{\text{number of deaths associated with the cause}}{\text{total number of deaths}}$$

Example 1.1

A butadiene plastics manufacturing plant is located in Beaverville, and the atmosphere is seriously contaminated by butadiene, a suspected carcinogen. The cancer death rate in the community of 8000 is 36 people per year and the total death rate is 106 people per year. Does Beaverville appear to be a healthy place to live?

The overall annual cancer death rate in the United States (from Table 1–1) is 193 death/10^5 population. The normal death rate in the United States is 2.2×10^6/year for a population of 240×10^6; thus the expected death rate for all causes for a community of 8000 is 73 deaths annually. Calculating risk in all three ways:

1.   Risk of death (SMR) $= \dfrac{106}{73} = 1.45$

2.   Risk of death $= \dfrac{36}{8000} = 450 \dfrac{\text{deaths}}{100{,}000}$

   (compared to the expected 193)

3.   Risk of death due to exposure $= \dfrac{36}{106} = 0.3 \dfrac{\text{annual cancer deaths}}{\text{total annual deaths}}$

   (compared to the expected 0.22)

In summary:

1.   Risk of dying from cancer is 1.45 greater in Beaverville than it is on average in the rest of the United States.
2.   Risk of dying of cancer *in any given year* is 450/193 = 2.33 times greater in Beaverville than in the United States as a whole.
3.   Risk of cancer being the cause of death is 0.30/0.22 = 1.54 times greater in Beaverville than on average in the rest of the United States.

Note again that in the above example it is *not* possible to say that the butadiene contaminated atmosphere *is the cause* of what clearly appears to be an unhealthy environment. The only valid conclusion is that living in Beaverville is relatively unhealthy because one is more likely to die of cancer. There could be other causes, however, such as a restricted diet of pizza and beer or a propensity to spend long hours in smoky bowling alleys. To repeat: A calculation of risk is *not* a determination of cause.

It seems clear that meaningful risk analyses can be done only with very large populations. Moreover, the health risk posed by most pollutants is observed to be considerably lower than the risks cited in Table 1–1 and may simply not be observed in a small population. Chapter 17 cites several examples of statistically valid risks from air pollutants.

The Environmental Protection Agency (EPA) has adopted the concept of *unit risk* in discussions of potential risk. Unit risk is the risk to an individual from exposure to a concentration of 1 g/m^3 of an airborne pollutant, or 10^{-9} g/L of a waterborne pollutant. The unit risk concept uses the second definition of risk: what the chances are of dying from a particular pollutant-related cause as compared to all other ways of dying. *Unit lifetime risk* is the risk to an individual from exposure to the above concentration for 70 years, a lifetime, while *unit occupational risk* implies exposure for 40 hours per week for 50 years, a working lifetime.

Example 1.2

EPA has calculated that the unit lifetime risk from exposure to ethylene dibromide (EDB) in drinking water is 0.85×10^{-5}. What is the lifetime risk to an individual who for five years drinks water with an average EDB concentration of 5g/L?

$$Risk = \frac{(concentration)(unit\ risk)(exposure\ time)}{70\ yr.}$$

$$Risk = (5 \times 10^{-12}g/L)\frac{0.85 \times 10^{-5}}{1 \times 10^{-9}g/L}(5yr/70yr) = 3.0 \times 10^{-9}$$

or for a population exposed to this risk, approximately 3 deaths in a billion.

Note that if either the concentration or the time of exposure had been zero the risk would also have been zero. But any finite level of a pollutant and any exposure time would lead to a risk as long as the unit risk is greater than zero.

Finally, there is a tendency to classify risk as voluntary or involuntary. People who smoke cigarettes, for example, subject themselves to great voluntary risk. Yet these same people are often adamantly opposed to a solid waste incinerator for their community because the risk, however miniscule, is nevertheless involuntary. Research has shown that people are generally willing to accept involuntary risks that are at least 1000 times less than voluntary risks. For example, there are 55,000 traffic fatalities in the United States annually, or a risk of $55,000/240 \times 10^6 = 230 \times 10^{-6}$ or 1 chance in 4300, of getting killed on the highways during any given year, using the second definition of risk. This risk is accepted voluntarily by a person every time he or she rides in a car.

The construction of a chemical processing plant is usually acceptable to the general public if the involuntary risk of death from accidental discharge is less than 1 in 1 million, or at least 1000 times less than riding on the highway.

THE "LULU" AND THE "NIMBY"

Since the passage of the Resource Conservation and Recovery Act (RCRA) in 1976 and particularly after the accident at the Three Mile Island nuclear plant in 1979, general awareness of the threats posed to human health by toxic or polluting substances in the environment has increased markedly. Some members of the public appear to be unwilling to accept *any* involuntary risk in their immediate environment. It has thus become increasingly difficult to find locations for facilities that can be suspected of producing any toxic or hazardous effluent: municipal land-fills; hazardous or radioactive waste sites; sewage treatment plants; incinerators; and even facilities like prisons, mental hospitals, or military installations whose lack of desirability is social rather than environmental. Siting local undesirable land uses, or LULUs as Frank Popper has termed them,[2] has become a dominant focus of environmental concern.

Local opposition to LULUs frequently can be summed up in the words "not in my back yard" (and local opponents of a particular LULU are often referred to by the acronym for this phrase, NIMBY). The environmental engineer is cautioned to identify the fine line between real concern about environmental degradation and an almost automatic "not in my back yard" reaction. If the latter is couched in terms of unwillingness to accept involuntary risk, it may be impossible to tell the difference. The environmental engineer recognizes (as many people do not) that virtually all human activity entails some—voluntary or involuntary—risk, that a risk-free environment is impossible to achieve. Often an environment that appears less risky has only traded one risk for another. The balance between risk and benefit to various segments of the population impinges directly on an emerging aspect of environmental ethics.

Is it ethical to site an undesirable facility where there is less local objection (perhaps because the employment is needed) instead of in the environment where it will do the least damage? One consequence of pollution control legislation in the United States has been the siting of U.S.-owned plants with hazardous or toxic effluents in countries that have little or no pollution control legislation (for example, oil desulfurization in Venezuela and copper smelting in the Philippines). The ethics of this phenomenon, sometimes called "pollution export," deserve closer examination than they have had.

Risk of death (mortality risk) is easier to determine for populations in the developed countries than is risk of illness (morbidity) because all deaths and their apparent causes are reported. Death certificate data can still be misleading: An individual who suffers from high blood pressure but is killed in an automobile accident becomes an accident statistic rather than a cardiovascular disease statistic. In the United States until very recently, statistics for deaths caused by occupational exposure could be determined only for males because the majority of women did not work outside the home for much of their adult lives.

---

[2] Frank J. Popper, "The Environmentalist and the LULU," *Environment* 27, no. 2 (1985): 7–40.

These particular uncertainties can be overcome in assessing risk from a particular cause by isolating the influence of that cause. This requires studying two populations whose environment is virtually identical except that the risk to be studied is present in one population and not in the other. Such a study is called a *cohort study,* and can be used to determine morbidity as well as mortality risk. One cohort study, for example, showed that residents of copper smelting communities who were exposed to airborne arsenic had a higher incidence of a certain type of lung cancer than did residents of similar industrial communities where there was no airborne arsenic.

Retrospective cohort studies are almost impossible to perform because of uncertainties in data on habits, other exposures, and so forth. Cohorts must be well matched in cohort size, age distribution, lifestyle, and other environmental exposures and must be large enough for an effect to be distinguishable from the deaths or illnesses that occur anyway. And finally even cohort studies do not *prove* anything with respect to cause and effect.

EXPOSURE AND LATENCY

Many cancers grow very slowly and are found (expressed) many years after the exposure to the responsible carcinogen. The length of time between exposure to a pollutant and expression of the adverse effect is called the *latency period.* Malignant neoplasms and leukemia occurring in adults have apparent latency periods ranging in length from about 10–40 years. Relating a cancer to a particular exposure is fraught with inherent inaccuracy: It is exceedingly difficult to isolate the effect of a single carcinogen when examining 30 or 40 years of a person's life. Thus many carcinogenic effects are simply unobserved.

There are a few instances in which a particular neoplasm is found only on exposure to a given agent (for example, a certain type of hemangioma is found only on exposure to vinyl chloride monomer) but for most cases the connection between exposure and effect is far from clear. Many carcinogens are identified through animal studies, but one cannot always extrapolate from animal results to human results. Finally, chronic low-level exposure may (or may not) have different effects than acute high-level exposure even when the total dose is the same. The cumulative uncertainty surrounding the epidemiology of pollutants has resulted in a conservative posture toward regulation and control. That is if there is any evidence, even inconclusive evidence, that exposure to a substance results in adverse health effects, release of that substance into the environment is regulated and controlled.

DOSE-RESPONSE EVALUATION

The effect of a pollutant—an organism's response to the pollutant—always depends in some way on the amount or dose of the pollutant to the organism. The magnitude of the dose in turn depends on the pathway into the organism. Pollutants have different effects depending on whether they are inhaled, ingested, or absorbed through the skin or whether exposure is external. Ingestion or inhalation determine the biochemical pathway of the pollutant in the organism. In general, the human body detoxifies an ingested pollutant more efficiently than it does an inhaled pollutant.

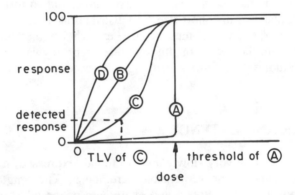

Figure 1-1   Dose-response curve.

The relationship between the dose of a pollutant and the organism's response can be expressed in a dose-response curve as shown in Figure 1-1. Some characteristic features of the dose-response relationship are:

1.   *Threshold.* The existence of a threshold in health effects of pollutants has been debated for many years. With reference to Figure 1-1, there are four basic dose-response curves possible for a dose of a specific pollutant (for example, carbon monoxide) and the response (for example, reduction in the blood's oxygen-carrying capacity). Curve A shows that if this dose-response relationship holds there is no effect on human metabolism until a critical concentration is reached. This concentration is called the *threshold* as is indicated in the figure.

     Curve B suggests that there is a detectable response for *any* finite concentration of the pollutant; this curve is a linear dose-response curve with no threshold.

     Curve C is a sigmoidal dose-response curve and is characteristic of many pollutant dose-response relationships. Although Curve C has no clearly defined threshold, the point at which a response can be detected (shown in the figure) is called the *threshold limit value (TLV)*. Occupational exposure guidelines are frequently set at the TLV. Curve C is sometimes called a *sublinear dose-response relationship*. Curve D represents a *supralinear relationship*, which is found when low doses appear to provoke a disproportionately large response.

2. *Total body burden.* Not all of our dose of pollutants comes from air. For example, although we breathe in about 50 μg/day of lead, we take in about 300 μg/day of lead in our water and food. In the setting of air quality standards for lead, it must therefore be recognized that most of the lead intake is from food and water. The sum of the organism from all pathways is the *total body burden* and includes concentrations of pollutants remaining in the body from previous exposures. The rate at which pollutants are eliminated from the body is measured by the physiological halflife.

3. *Time versus dosage.* Most pollutants require time to react and thus the time of contact is as important as is the level. The best example of this is the effect of carbon monoxide. CO reduces the oxygen-carrying capacity of the blood by combining with the hemoglobin and forming carboxyhemoglobin. At about 60 percent carboxyhemoglobin concentration death results from lack of oxygen. The effects of CO at sublethal concentrations are usually reversible. Because of the time-response problem, ambient air quality standards are set at maximum allowable concentrations for a given time (Chapter 21).

4. *Synergism.* This is defined as a combined effect that is greater than the sum of individual effects. For example, black lung disease in coal miners occurs only when the miner is also a cigarette smoker. Coal mining by itself or cigarette smoking by itself will not cause black lung, but the synergistic action of the two puts miners who smoke at high risk. The opposite of synergism is *antagonism,* a phenomenon that occurs when two pollutants counteract each other's effects.

5. *LC_{50} and LD_{50}.* Dose-response relationships for people are generally determined from health data or epidemiological studies. Human volunteers obviously cannot be subjected to doses of pollutants that produce major or lasting health effects. Toxicity can be determined, however, by subjecting nonhuman organisms to increasing doses of pollutants until the organism dies. The term LD_{50} refers to the dose which is lethal for 50 percent of the experimental animals used. (LC_{50} refers to lethal concentration instead of lethal dose.) LD_{50} values are of most use in comparing toxicities, as for pesticides and agricultural chemicals; no direct extrapolation is possible, either to humans or to any species other than the one used for the LD_{50} determination. LD_{50} for people can sometimes be determined retrospectively when a large population has been exposed accidentally, as in the accident at the Chernobyl nuclear reactor.[3]

6. *Bioaccumulation and bioconcentration.* The term *bioaccumulation* is used when a substance is concentrated in one organ or type of tissue.

[3] U.S. Department of Energy, *Health and Environmental Consequences of the Chernobyl Nuclear Power Plant Accident* DOE/ER–0332, 1987, p. 2.3.

Iodine, for example, bioaccumulates in the thyroid gland. The dose to an organ can thus be considerably greater than the body burden, or whole-body, dose.

Bioconcentration occurs with movement up the food chain. A study[4] of a Lake Michigan ecosystem found the following bioconcentration of DDT:

0.014 ppm (wet weight) in bottom sediments

0.41 ppm in bottom-feeding crustacea

3 to 6 ppm in fish

2400 ppm in fish-eating birds

POPULATION RESPONSES

Individual responses to pollutants are not identical. Dose-response curves will differ from one person to another; in particular, thresholds will differ. In general, threshold values in a population follow a Gaussian distribution.

Individual responses and thresholds also depend on age, sex, general state of health, and so forth. As might be anticipated, healthy young adults are less sensitive to pollutants than are elderly people, chronically or acutely ill people, and children. Allowable releases of pollutants are in theory restricted to amounts that assure protection of the health of the entire population including its most sensitive members. In many cases, however, such restriction would mean zero release.

The levels actually chosen take technical and economical control feasibility into account and usually are set below threshold level for 95 percent or more of the entire population. For nonthreshold pollutants, however, no such determination can be made. In these instances, there is no release level for which protection can be assured for anyone so that comparative risk analysis is necessary.

Many pollutants, including carcinogens, are considered to have no measurable threshold. The best available control for such pollutants still entails a residual risk. There is a continuing need in our industrial society therefore for accurate quantitative risk assessment. We must also remain aware that we have not precisely identified carcinogens or mutagens, nor do we understand their mechanism of action. We can only identify apparent associations between most pollutants and a given health effect.

In almost all cases, doses to the general public are so small that excess mortality and morbidity are not identifiable. In fact, almost all of our knowledge of adverse health effects or pollutants comes from occupational exposure, where doses are orders of magnitude higher.

---

[4] J.J. Hickey *et al.*, *Jour. Appl. Ecol.* 3 (1966): 141–154.

CONCLUSION

Risk due to the contamination of our environment is a ubiquitous and inevitable part of our lives. Risks can happily be reduced significantly by controlling environmental pollution. The philosophy, regulatory approaches, and engineering design of environmental pollution control comprise the remainder of this book.

PROBLEMS

1.1 Workers in a chemical plant producing molded polyvinyl chloride plastics suffered from hemangioma, a form of liver cancer that is usually fatal. It was found that during the 20 years of the plant's operation, 20 employees out of 350—the total number of employees at the plant during those years—developed hemangioma. Does working in the plant present an excess cancer risk? Why? What assumptions need to be made?

1.2 Additional, previously unavailable, data on hemangioma incidence in the general population is developed. It is found that among people who have never worked in the plastics or chemical industry there are only 10 deaths per 10^5 persons per year from hemangioma. How does this change your answer to Problem 1?

1.3 The unit lifetime risk from airborne arsenic is 9.2×10^{-3}. EPA regards an acceptable annual risk from any single source to be 10^{-7}. A copper smelter emits arsenic into the air, and the average concentration in the air within a two-mile radius of the smelter is 5.5 $\mu g/m^3$. Is the risk from smelter arsenic emissions acceptable to EPA?

1.4 A group of arthritis sufferers were asked to gauge the ability of a new drug, Nopain, to relieve their arthritis pain. They were asked to rank the pain on a scale of 0 to 5 where a rank of 5 meant no relief and a rank of 0 meant complete relief. The average rankings developed for various doses of Nopain are given in the table below.

Rank	Dose of Nopain
5.0	30 mg
5.0	100 mg
4.8	150 mg
3.0	200 mg
1.2	250 mg
0.7	300 mg
0.5	350 mg
0.5	400 mg

Draw a dose-response curve, and indicate the threshold limit value.

1.5 Plutonium has a physiological halflife of one year. If a volunteer eats 5.0 mg of plutonium, how much is left in his body after 3 years?

1.6 By what factor is DDT bioconcentrated in birds from the fish the birds eat? From the bottom sediments?

Chapter 2
Water Pollution

Although people intuitively relate filth to disease, the fact that pathogenic organisms can be transmitted by polluted water was not recognized until the middle of the nineteenth century. Probably the most dramatic demonstration that water can indeed transmit disease was the Broad Street pump-handle incident.

A public health physician named John Snow, assigned to attempt to control the 1852 cholera epidemic, realized that there seemed to be an extremely curious concentration of cholera cases in one part of London. Almost all of the people affected drew their drinking water from a community pump in the middle of Broad Street. Even more curious was the fact that the people who worked and lived in an adjacent brewery were not afflicted. Although this seemed to demonstrate the health benefits of beer, welcome news to most students, Snow recognized that the absence of cholera in the brewery might be because the brewery obtained its water from a private well and not the Broad Street pump.

Snow's evidence convinced the city council to ban the obviously polluted water supply, which was done by simply removing the pump handle, thus effectively preventing the people from using the water. The source of infection was stopped, the epidemic subsided, and a new era of public health awareness related to water supplies began.

The concern with water pollution was, until recently, a concern about health effects. In many countries it still is. In the United States and other developed countries, however, water treatment and distribution methods have for the most part eradicated the transmission of bacterial waterborne disease. We now think of water pollution not so much in terms of health as in terms of its effects on the aquatic ecosystem and on fish and shellfish in particular. We also consider conservation, aesthetics, and the preservation of natural beauty and resources. People have an inexplicable affinity

15

for water, and the fouling of lakes, rivers and oceans is intrinsically unacceptable to the concerned citizen.

SOURCES OF WATER POLLUTION

Industries, municipalities, and stormwater runoff contribute to the pollution of natural water systems. The United States has more than 40,000 factories that use water, and their industrial wastes are probably the greatest single water pollution problem. Industry is creating a wide array of new chemicals each year all of which can eventually find their way into the water. Decomposition of these chemicals and how they react with each other is understood only poorly, as are their acute and chronic toxicity.

Another industrial waste is heat. Heated discharges can drastically alter the ecology of a stream or lake. This alteration is sometimes called beneficial, perhaps because of better fishing or an ice-free docking area. The deleterious effects of heat, in addition to promoting modifications of ecological systems, include a lessening of dissolved oxygen solubility and increases in metabolic activity. Dissolved oxygen is vital to healthy aquatic communities, and the warmer the water, the more difficult it is to get oxygen into solution. As the temperature increases the metabolic activity of aerobic (oxygen-using) aquatic species increases, thus demanding more oxygen. It is a small wonder, therefore, that the vast majority of fish kills due to oxygen depletion occur in the summer.

Municipal waste is a source of water pollution second in importance only to industrial wastes. In the United States more than 20,000 municipal wastewater facilities discharge into rivers and streams. Around the turn of the century, most discharges from municipalities received no treatment whatsoever. In the United States, sewage from 24 million people was flowing directly into our watercourses. Since that time, the population has increased and so has the contribution from municipal discharges. It is estimated that presently the *population equivalent*[1] of municipal discharges to watercourses is about 100 million. Even with the billions of dollars spent on building wastewater treatment plants, the contribution from municipal pollution sources has not been significantly reduced. We

---

[1]Population equivalent is the number of people needed to contribute a certain amount of pollution. For example, if a town has 10,000 people, and the treatment plant is 50 percent effective, their discharge has a population equivalent of 5000 people. Similarly, if an industry discharges 1000 lb of solids per day and if each person contributes 0.2 lb per day into domestic wastewater, the industrial waste can be expressed as being the equivalent of $1000/0.2 = 5000$ people.

seem to be holding our own, however, and at least are not falling further behind.

One problem, especially in the older cities on the east coast of the United States, is the sewerage systems. When the cities were first built, the engineers realized that sewers were necessary for both stormwater and sanitary wastes, and they saw no reason why both stormwater and sanitary wastes should not flow in the same pipelines. After all, they both ended up in the same river or lake. Such sewers are now known as *combined sewers*.

As years passed and populations increased, the need for the treatment of sanitary wastes became obvious, and two-sewer systems were built, one to carry stormwater and the other, sanitary waste. Such systems are known as *separate sewers*.

Almost all of the cities with combined sewers have built treatment plants that can treat the *dry weather flow* (sanitary wastes). As long as it doesn't rain, they can provide sufficient treatment. When the rains come, however, the flows swell to many times the dry weather flow and most of it must be bypassed directly into a river or lake. This overflow contains sewage as well as stormwater and has a high polluting capacity. All attempts to capture this excess flow for subsequent treatment, such as storage in underground caverns and rubber balloons, are expensive. The alternative solution, however, separating the sewers, is extremely expensive.

In addition to industrial and municipal wastes, water pollution emanates from many "non-point" sources.

Agricultural wastes, should they all flow into streams, would have a population equivalent of about 2 billion. The problems are intensifying with the increase in the size and number of feedlots, cattle pens constructed for the purpose of fattening the cattle before slaughter. These feedlots are usually close to slaughterhouses (hence cities) and a large number of animals are packed into a small space. Drainage for these lots has an extremely high pollutional strength, as does drainage from intensive duck and chicken cultivation. There is a high potential for water pollution whenever animal waste is concentrated, even in aquaculture.

Sediment from land erosion can also be classified as a pollutant. Sediment consists of mostly inorganic material washed into a stream as a result of farming, construction or mining operations. The detrimental effects of sediment include interference with the spawning of fish by covering gravel beds, interference with light penetration thus making food more difficult to find, and direct damage to gill structures. In the long run, sediment could well be one of our most harmful pollutants.

The concern with pollution from petroleum compounds is relatively new, starting to a large extent with the Torrey Canyon disaster in 1967. Ignoring maps showing submerged rocks, the huge tanker loaded with crude oil plowed into a reef in the English Channel. Almost immediately oil began seeping out, and both the French and British became concerned. Rescue efforts failed and the Royal Air Force attempted to set it on fire, with little success. Almost all of the oil eventually leaked out and splashed on the beaches of France and England. The French started the back-breaking chore of spreading straw on the beaches, allowing the straw to adsorb the oil, and then collecting and burning the oil-soaked straw. The English used detergents to disperse the oil and then flushed the emulsion off the beaches. Time has shown the French way to be best since the English detergents have now been shown to be potentially more harmful to coastal ecology than the oil would have been.

Although the Torrey Canyon disaster was the first big spill, many have followed it. It is estimated that there are no fewer than 10,000 serious oil spills in the United States every year. In the spring of 1989, a grounded tanker in Alaska's Gulf of Valdez spilled enough oil that complete cleanup proved impossible. In addition, the contribution from routine operations such as flushing oil tankers may well exceed all the oil spills.

The acute effect of oil on birds, fish, and microorganisms is reasonably well cataloged. What is not so well understood and potentially more harmful is the subtle effect on other aquatic life. Anadromous fish, for example, have been known to find their home stream by the specific smell (or taste) of the water, caused in large part by the hydrocarbons present. If people continue (albeit unintentionally) to pour hydrocarbons into salmon runs, it is possible that the salmon will become so confused that they will refuse to enter their spawning stream.

Another form of industrial pollution, much of it willed to us by our ancestors, is acid mine drainage. The problem is caused by the leaching of sulfur-laden water from old abandoned mines (as well as some active mines). On contact with air, these compounds are soon oxidized to sulfuric acid, a poison to all living matter.

It should be amply clear, therefore, that water can be polluted by many types of waste products.

Water pollution problems (and hence their solutions) can be best understood by first describing them in the context of an ecosystem, and then studying one specific aspect of that ecosystem—the biodegradation of organics.

ELEMENTS OF AQUATIC ECOLOGY

Plants and animals in their physical environment make up an *ecosystem*. The study of such ecosystems is *ecology*. Although we often draw lines around a specific ecosystem in order to be able to study it more fully (for example, a farm pond) and in so doing assume that the system is totally self-contained, this obviously is not true, and we must remember that one of the tenets of ecology is that "everything is connected with everything else."

Within an ecosystem there exist three broad categories of actors. The *producers* take energy from the sun and nutrients such as nitrogen and phosphorus from the soil and through the process of *photosynthesis* produce high-energy chemicals. The energy from the sun is thus stored in the chemical structure of their organic molecules. These organisms are often referred to as *autotrophs*.

A second group of organisms are the *consumers* who use some of this energy by ingesting the high-energy molecules. These organisms are in the second trophic level in that they directly use the energy of the producers. There can be several more trophic levels as the consumers use the level above as a source of energy. A simplified ecosystem showing various trophic levels is shown in Figure 2–1, which also illustrates the progressive use of energy through the trophic levels.

The third group of organisms, the *decomposers* or decay organisms, use the energy in animal wastes and dead plants and animals, and in so doing convert the organic molecules to stable inorganic compounds. The residual inorganics then become the building blocks for new life, using the sun as the source of energy.

Ecosystems exhibit a flow of both energy and nutrients. Energy flow is in only one direction: from the sun and through each trophic level. Nutrient flow, on the other hand, is cyclic. Nutrients are used by plants to make high-energy molecules, which are eventually decomposed to the original inorganic nutrients, ready to be used again.

The entire food web, or ecosystem, stays in dynamic balance, with adjustments being made as required. Such a balance is called *homeostasis*. For example, a drought one year may produce little grass thus exposing field mice to predators such as owls. The mice, in turn, spend more time in burrows thus not eating as much and allowing the grass to reseed for the following year. External perturbations, however, can upset and destroy an ecosystem. In the previous example, the use of an herbicide to kill the grass might also destroy the field mouse population, and in turn diminish the number of owls. It must be recognized that although most ecosystems can absorb a certain amount of insult, a sufficiently large perturbation can cause irreparable damage.

The amount of perturbation a system is able to absorb without being destroyed is tied to the concept of the *ecological niche*. The combination of function and habitat of an organism in an ecological system is its niche. A niche is not a property of a type of organism or species but is its best accommodation with the environment. In the example above, the grass is

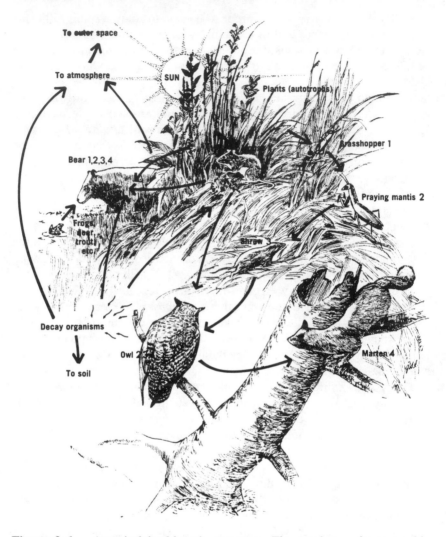

Figure 2–1   A typical land-based ecosystem. The numbers refer to trophic level, and the arrows show progressive loss of energy for life systems [From Turk, A. et al. *Environmental Science* (Philadelphia: W. B. Saunders Co., 1974).]

a producer that acts as food for the field mouse, which in turn is food for the owl. If the grass were destroyed, both the mice and owls might eventually die out. But suppose there were *two* types of grass, each equally acceptable as mouse food. Now if one died out, the ecosystem would not be destroyed because the mice would still have food. This simple example demonstrates an important ecological principle: the stability of an ecosystem is proportional to the number of organisms capable of filling various niches. A jungle, for example, is a very stable ecosystem, whereas the tundra in Alaska is extremely fragile. Another fragile system is the deep ocean—a fact which should be a consideration in the disposal of hazardous and toxic materials in deep ocean areas. Inland water courses tend to be fairly stable ecosystems but certainly not totally resistant to destruction by outside forces. Other than the direct effect of toxic materials such as heavy metals and refractory organics,[2] the most serious effect of water pollution for inland waters is the depletion of dissolved (free) oxygen. All higher forms of aquatic life exist only in the presence of oxygen, and most desirable microbiologic life also requires oxygen. Generally, all natural streams and lakes are *aerobic* (containing dissolved oxygen). If a watercourse becomes *anaerobic* (absent of oxygen), the entire ecology changes to make the water unpleasant or unsafe.

Problems associated with pollutants that affect the dissolved oxygen levels cannot be appreciated without a fuller understanding of the concept of decomposition or biodegradation, part of the total energy transfer system of life.

BIODEGRADATION

Plant growth, or photosynthesis, can be represented by the equation

$$CO_2 + H_2O \xrightarrow[\text{\& nutrients}]{\text{sunlight}} HCOH + O_2$$

In this representation formaldehyde (HCOH) and oxygen are produced from carbon dioxide and water with sunlight the source of energy.[3] If the formaldehyde and oxygen are combined and ignited, an explosion results. The energy released during such an explosion is stored in the carbon-hydrogen-oxygen bonds of formaldehyde.

As discussed above, plants (producers) use inorganic chemicals as nutrients and, with sunlight as a source of energy, build high-energy molecules. The animals (consumers) eat these high-energy molecules, and during their digestion process some of the energy is released and used by the animals. The release of this energy is quite rapid and the end products of digestion (excrement) consist of partially stable compounds. These

---

[2]Refractory organics are manufactured organic materials, such as the pesticide DDT, which decompose very slowly in the environment.

[3]Of course, formaldehyde is not the end product of photosynthesis, but it is an organic molecule and happens to provide a simple equation.

compounds become food for other organisms and are thus degraded further but at a slower rate. After several such steps, very low-energy compounds are formed that can no longer be used by microorganisms for food. Plants then use these compounds to build more high-energy molecules and the process starts all over. The process is symbolically shown in Figure 2–2.

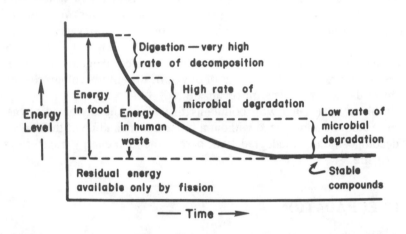

Figure 2–2   Energy loss in biodegradation. [After P. H. McGauhey, *Engineering Management of Water Quality* (New York: McGraw-Hill, 1968).]

It is important to realize that many of the organic materials responsible for water pollution enter watercourses at a high energy level. It is the biodegradation, or the gradual use of this energy, by a chain of organisms that causes many of the water pollution problems.

AEROBIC AND ANAEROBIC DECOMPOSITION

Decomposition, or biodegradation, can take place in one of two distinctly different ways: aerobic (using free oxygen) or anaerobic (in the absence of free oxygen).

The basic equation of aerobic decomposition is

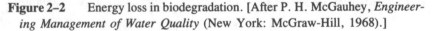

Complex Organics $+ O_2 \rightarrow CO_2 + H_2O +$ Stable Products

Carbon dioxide and water are always two of the end products of aerobic decomposition. Both are stable, low in energy, and used by plants in the process of photosynthesis. If sulfur compounds are involved in the reaction, the most stable end product is $SO_4^=$, the sulfate ion. Similarly, phosphorus ends up as $PO_4^=$, orthophosphate. Nitrogen goes through a series of increasingly stable compounds, finally ending up as nitrate. The progression is

Organic Nitrogen $\rightarrow$ NH_3 (ammonia) $\rightarrow$ NO_2^- (nitrite) $\rightarrow$ NO_1^- (nitrate)

Because of this distinctive progression, nitrogen has been in the past and to some extent is still used as an indicator of pollution.

A schematic representation of the aerobic cycle for carbon, sulfur and nitrogen compounds is shown as Figure 2–3. This figure illustrates only the basic facts, and is a gross simplification of the actual steps and mechanisms involved.

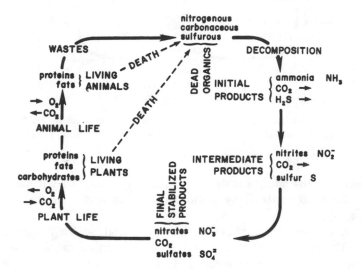

Figure 2–3 Aerobic nitrogen, carbon and sulfur cycles. [After P. H. McGauhey, *Engineering Management of Water Quality* (New York: McGraw-Hill, 1968).]

A second type of biodegradation is anaerobic, performed by a completely different set of microorganisms to which oxygen is in fact toxic. The basic equation of anaerobic decomposition is

Complex Organics $\rightarrow CO_2 + CH_4 +$ other partially stable compounds

Note that many of the end products shown are biologically unstable. CH_4, for example, is methane, a high-energy gas commonly called marsh gas, physically stable but still able to be decomposed biologically. Nitrogen compounds stabilize to ammonia (NH_3), and sulfur ends up as evil-smelling hydrogen sulfide (H_2S) gas. Figure 2–4 is a schematic

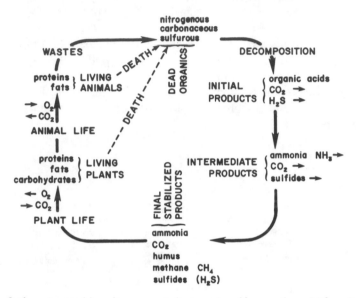

Figure 2–4   Anaerobic nitrogen, carbon and sulfur cycles. [After P. H. McGauhey, *Engineering Management of Water Quality* (New York: McGraw-Hill, 1968).]

representation of anaerobic decomposition. Note that the left half of the cycle, the photosynthesis by plants, is identical to the aerobic cycle in Figure 2–3.

Biologists often speak about various compounds as "hydrogen acceptors." The hydrogen atoms, torn from high-energy organic molecules, must be attached to various compounds. In aerobic decomposition oxygen serves this purpose and is thus known as the hydrogen acceptor. It accepts the hydrogen atoms to form water.

In anaerobic decomposition free oxygen is not available, and the next preferred hydrogen acceptor is nitrogen, thus forming ammonia, NH_3. If free oxygen is not available, ammonia cannot be converted to nitrites or nitrates. If nitrogen is not available, the next preferred hydrogen acceptor is sulfur, thus forming hydrogen sulfide, H_2S, the chemical responsible for the notorious rotten egg smell.

EFFECT OF POLLUTION ON STREAMS

When a high-energy organic material such as raw sewage is discharged to a stream, a number of changes occur downstream from the point of discharge. As the organics are decomposed, oxygen is used at a greater rate than before the pollution occurred, and the dissolved oxygen (DO) level drops. The rate of reaeration, or solution of oxygen from the air, also increases, but this is often not great enough to prevent a total depletion of oxygen in the stream. When this happens, the stream is said to become anaerobic. Often, however, the DO does not drop to zero, and the stream recovers without experiencing a period of anaerobics. The latter situation is depicted graphically in Figure 2–5. The dip in DO is referred to as a *dissolved oxygen sag curve.*

The relationship shown in Figure 2–5 is the result of a dynamic balance between how much oxygen the stream can take in from the atmosphere

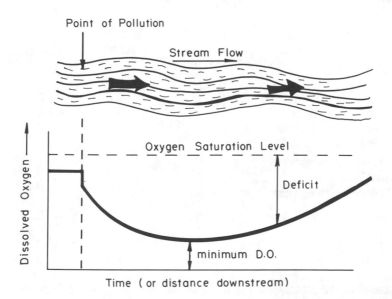

Figure 2–5   Dissolved oxygen downstream from a source of organic pollution. The curve depicts a DO sag without anaerobic conditions.

and how much oxygen is being used by the microorganisms. If the rate of use is great, as it is in the beginning of the polluted stream, the DO level drops rapidly since the supply rate from the atmosphere cannot keep up with it. As the rate of oxygen use decreases, the supply begins to be able to keep pace, and the curve reaches a minimum. If no further pollution is added, the DO sag curve eventually recovers and the DO levels once again reach saturation.

Stream flow is of course variable, and the critical dissolved oxygen levels can be expected to occur when the flow is the lowest. Accordingly, most state regulatory agencies base their calculations on a statistical low flow, such as a 7-day, 10-year low flow or the seven consecutive days of lowest flow that can be expected to occur once during a 10-year interval. This is calculated by first estimating the lowest 7-day discharge for each year, assigning these rankings, m as $m = 1$ for the most severe (least flow) and $m = n$ (where n is the number of years) for the least severe case. The probability of a flow equal to or more than a low flow occurring is calculated as $m/(n + 1)$, and plotted versus the flow. Usually log-probability paper is used since this gives the best straight line fit. Then $m/(n + 1) = 0.1$ is read as the 10-year low flow.

Example 2.1

Calculate the 10-year, 7-day low flow given the data below.

Year	Lowest flow 7 consecutive days (m^3/sec)	Ranking (m)	Lowest flow in order of severity (m^3/sec)	$\dfrac{m}{n + 1}$
1965	1.2	1	0.4	0.071
1966	1.3	2	0.6	0.143
1967	0.8	3	0.6	0.214
1968	1.4	4	0.8	0.285
1969	0.6	5	0.8	0.357
1970	0.4	6	0.8	0.428
1971	0.8	7	0.9	0.500
1972	1.4	8	1.0	0.571
1973	1.2	9	1.2	0.642
1974	1.0	10	1.2	0.714
1975	0.6	11	1.3	0.785
1976	0.8	12	1.4	0.857
1977	0.9	13	1.4	0.928

These data are plotted on Figure 2–6 and the minimum 7-day, 10-year flow is read off as 0.5 m³/sec.

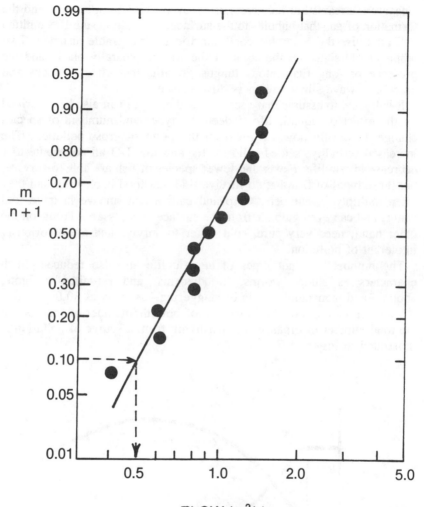

Figure 2–6   Plot of 10-year, 7-day low flows for Example 2.1

When the rate of oxygen usage overwhelms the rate at which the oxygen can be supplied, the stream may go anaerobic. An anaerobic stream is easily identifiable. Since oxygen is no longer around to act as the hydrogen acceptor, ammonia and hydrogen sulfide are formed among other gases. Some of these gases will dissolve readily, but others will attach themselves as bubbles to hunks of black solid material known as *benthic deposits* (or simply sludge), and buoy this material to the surface.

Anaerobic streams are thus recognized by floating sludge solids and the formation of gas that bubbles to the surface. In addition, the H_2S emitted will advertise the anaerobic condition for a considerable distance. Two other telltale signs are the color of the water, generally black, and the presence of long filamentous fungus growths that cling to rocks and gracefully wave slimy streamers downstream.

It is logical to assume that such outward changes can also be described by the effect on aquatic life. Indeed the types and numbers of species change drastically downstream from the point of gross pollution. The increased turbidity, settled solid matter and low DO all contribute to a decrease in fish life. Fewer and fewer species of fish are able to survive, but those types of fish that do survive find that food is plentiful and they often multiply in numbers. Carp and catfish can survive in quite foul water, and can even gulp air from the surface if necessary. Trout, on the other hand, need very pure, cold water to survive and are notoriously intolerant of pollution.

The number of other types of aquatic life are also reduced. Such characters as sludge worms, bloodworms, and rat-tailed maggots, abound, and their numbers can be staggering—as many as 50,000 sludge worms per square foot. The variation of both the number of species and the total number of organisms downstream from a source of pollution is illustrated in Figure 2–7.

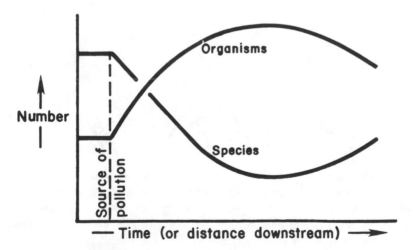

Figure 2–7   The number of species and the total number of organisms down-stream from a point of organic pollution.

Table 2–1.   Diversity of Aquatic Organisms: Sample Calculations

Location	Diversity Index ($\bar{d}$)
Above the Outfall	2.75
Immediately Below the Outfall	0.94
Downstream	2.43
Further Downstream	3.80

The diversity of species can be quantified by using an index such as

$$\bar{d} = \sum_{i=1}^{5} \left(\frac{n_i}{n}\right) \log_2 \left(\frac{n_i}{n}\right)$$

where $\bar{d}$ = diversity index
          n_i = number of individuals in the i th species
          n = total number of individuals in all S species.

In one study, the diversity index was calculated above and below a sewage outfall, and the results were as shown in Table 2–1.

It was mentioned earlier that nitrogen compounds can be used as indicators of pollution. The changes in the various forms of nitrogen are shown in Figure 2–8. The first transformation, both in aerobic and anaerobic

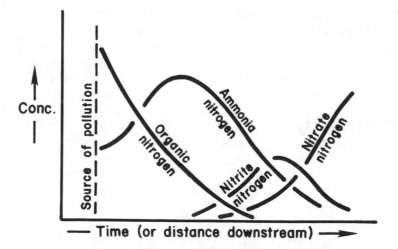

Figure 2–8   Typical variations in nitrogen compounds downstream from a point of organic pollution.

decomposition, is the formation of ammonia, and thus the concentration of ammonia increases as organic nitrogen decreases. Similarly, the concentration of nitrate nitrogen will finally increase to become by far the dominant form of nitrogen.

It is important to remember that the reactions of a stream to pollution outlined above occur when a rapidly decomposable organic material is wasted. The stream will react much differently to an inorganic waste, say from a metal-plating plant. If the waste is toxic to aquatic life, both the kind and total number of organisms will decrease below the outfall. The DO will not fall and might even rise. There are many types of pollution, and a stream will react differently to each. Even more complicated is a situation where two or more wastes are involved.

EFFECT OF POLLUTION ON LAKES

The effect of pollution on lakes differs in several respects from the effect on streams. For one thing, light and temperature have significant influences on a lake and must be included in any limnological[4] analysis. Light is the source of energy in the photosynthetic reaction, and the penetration of light into the lake water is important. This penetration is logarithmic. If, for example, at a 1-foot depth the light is 10,000 footcandles,[5] at 2 feet it might be 1000, at 3 feet 100, and at 4 feet only 10 footcandles. Only the top few feet of a lake generally experience light penetration, and hence all photosynthetic reactions occur in that zone.

Temperature often has a profound effect on a lake. Water is at a maximum density at 4 °C (water both colder and warmer is less dense; ice, therefore, floats). Water is also a poor conductor of heat and retains heat quite well.

Lakes usually undergo a seasonal variation in water temperature. These temperature-depth relationships are illustrated in Figure 2–9. During the winter, assuming the lake does not freeze, the temperature is often constant with depth. As warmer weather approaches the top layers begin to warm up. Since water is a poor conductor of heat and warmer water is lighter, a distinct temperature gradient is formed known as *thermal stratification*. These strata are often very stable and last through the summer months. The top layer is called the *epilimnion,* the middle the *metalimnion,* and the bottom the *hypolimnion.* The inflection point in the curve is called the *thermocline.* Circulation of water occurs only within a zone, and thus there is only limited transfer of biological or chemical material (including dissolved oxygen) across the boundaries.

---

[4]*Limnology* is the study of lakes.
[5]The intensity of light is measured in footcandles.

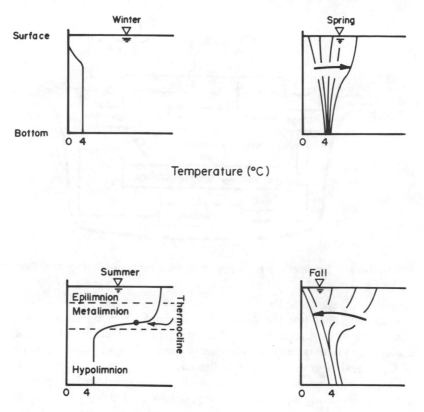

Figure 2–9 Typical temperature-depth relationships in a lake.

As the colder weather approaches, the top layers begin to cool, become more dense, and sink. This creates circulation throughout the lake, a condition known as *fall turnover*. Often a spring turnover also occurs.

The biochemical reactions in a natural lake can be represented schematically as in Figure 2–10. A river feeding the lake would contribute carbon, phosphorus, and nitrogen either as high-energy organics or as low-energy compounds. The *phytoplankton* or *algae* (microbial free-floating plants) take C, P, and N and, using sunlight as a source of energy, make high-energy compounds. Algae are eaten by *zooplankton* (tiny aquatic animals), which are in turn eaten by larger aquatic life such as fish. All of these forms of life defecate, thus contributing to a pool of *dissolved organic carbon*. This pool is further fed by the death of aquatic life. Bacteria utilize dissolved organic carbon and produce CO_2, in turn used by the algae. Additional CO_2 is provided from the respiration of the fish and zooplankton as well as the CO_2 dissolved directly from the air.

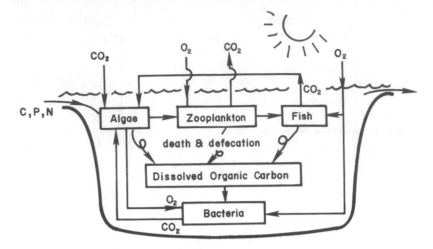

Figure 2–10 Schematic representation of lake ecology (Courtesy of Donald Francisco).

In an unpolluted lake the supply of incoming C, P, and N is sufficiently small to limit the production of algae, and the productivity of the entire ecological system is limited. But what happens when an excessive amount of C, P, and N are introduced to the lake?

The plentiful supply of nutrients promotes the uncontrolled growth of algae. When the algae, along with the zooplankton and fish, die, they drop to the bottom and become another source of carbon for the bacteria. Aerobic bacteria will use all available dissolved oxygen in decomposing this material and may use enough oxygen to deplete all available DO, thus creating anaerobic conditions.

As more and more algae are produced in the epilimnion, the bacteria in the lower portions of the lake will utilize more and more oxygen, and the metalimnion might also become anaerobic. All of the aerobic biological activity would thus concentrate in the upper few feet of the lake, the epilimnion. All this activity causes turbidity, thus decreasing light penetration, and in turn limiting algal activity to the surface layers. The amount of DO contributed by the algae is therefore decreased. Eventually, when the epilimnion is also anaerobic, all aerobic aquatic life disappears and the algae, because of limited light penetration, concentrate on the surface of the lake, forming large green mats or *algal blooms*. These algae will also die, and eventually fill up the lake, producing what we now know as a peat bog.

This entire process is called *eutrophication*. It is a continually occurring natural process. But what may have taken thousands of years to occur naturally can be accomplished in only a decade if enough nutrients are introduced into a lake as a result of human activities.

It is worth repeating that eutrophication is a natural process. The stages of a lake's natural aging are the *oligotrophic* stage during which both the variety and number of species grow rapidly, the *mesotrophic* stage during which there is a dynamic equilibrium of species in the lake, the *eutrophic* stage during which less complex organisms take over and the lake appears to become gradually choked with weeds. The addition of nutrients, however, can speed up eutrophication considerably. Addition of phosphorus is particularly effective, since phosphorus is often the limiting nutrient.[6]

Where do these nutrients originate? One source is excrement, since all human and animal wastes include C, P, and N. This source, however, is small compared to synthetic detergents and fertilizers. It is estimated that of the total P discharged to our lakes, 1/2 comes from agricultural runoff, 1/4 comes from detergents, and 1/4 from all other sources. It seems unfortunate that the presence of phosphates in detergents has received so much unfavorable attention when runoff from fertilized land is a much more important source. The conversion to nonphosphate detergents would be of limited value if other sources are not controlled.

It is generally believed that a P concentration between 0.01 and 0.1 mg/L is sufficient to promote accelerated eutrophication. Effluents from sewage treatment plants often contain from 5 to 10 mg/L of P. A river flowing through farm country might carry from 1 to 4 mg/L of P. In moving streams this high P concentration is not a problem since streams are continually flushed out and the algae do not have time to accumulate. Eutrophication can thus occur only in lakes, ponds, estuaries, and sometimes in very slowly moving rivers.

Incidentally, it has been estimated that only about 15 percent of the people in the United States should be concerned about phosphate discharges from their municipalities. About 85 percent of the wastewater flows into moving streams and rivers where eutrophication is not a problem.

There is also some question as to the validity of blaming phosphorus for accelerated eutrophication. Generally a P:N:C ratio of 1:16:100 is required for growth. It takes 16 parts N and 100 parts C for every part P for algae to grow. If there are more than 16 parts N and 100 parts C for every P, then N and C are said to be in excess and thus P must be limiting growth. Although this is the case with most lakes, there are those that have

---

[6]Phosphorus "limits" algal growth in that it is a constraint against unlimited growth, much like the gas pedal on your car "limits" the speed of your car. The car is able to go faster—all the components are there and could function at a higher rate—but the gas pedal is a constraint against higher speed. Dumping excess phosphorus into a lake is like tromping down on the gas pedal.

very high P levels and have few problems with algal blooms. Conversely, some lakes have experienced serious algal problems while carrying extremely low P levels. In addition, recent evidence suggests that nitrogen is limiting algal growth in brackish waters such as bays and estuaries. Suffice it to say the blame for the accelerated eutrophication of many of our lakes cannot at this time be placed on any one chemical or product. An interaction among the many pollutants involved is the probable answer.

HEAVY METALS AND TOXIC SUBSTANCES

In 1970, Barry Commoner[7] alerted the nation to the growing problem of mercury contamination of lakes, streams, and marine waters. Commoner identified the manufacture of chlorine and lye from brine—the chlor-alkali process—as a major source of mercury contamination. Elemental mercury is methylated by aquatic organisms, and methylated mercury finds its way into fish and shellfish and thus into the human food chain. Methyl mercury is a powerful neurological poison. Methyl mercury poisoning was first identified in Japan in the 1950s as "Minamata disease": Mercury-containing effluent from the Minamata Chemical Company was found to be the source of mercury in food fish.[8]

Arsenic, copper, lead, and cadmium are often deposited in lakes and streams from the air in the vicinity of nonferrous smelters. These substances are also constituents of mine drainage and industrial effluent. The effluent from electroplating and metal refining contains a number of heavy metal constituents. Heavy metals, copper in particular, can be toxic to fish as well as harmful to human health.

In the 1960s and 1970s, there were many incidents of surface water contamination in the United States with toxic and carcinogenic organic compounds. The sources of these include effluent from petrochemical industries and agricultural runoff, which contain both pesticide and fertilizer residues. Trace quantities of chlorinated hydrocarbon compounds in drinking water can also be attributed to the chlorination of organic residues by the chlorine that is added as a disinfectant.

EFFECT OF POLLUTION ON OCEANS

Not many years ago the oceans were considered infinite sinks; the immensity of the seas and oceans seemed impervious to assault. This is no longer

[7]Barry Commoner, *The Closing Circle,* New York, 1971.

[8]R. Hartung and B. Dinman, eds. *Environmental Mercury Contamination,* Ann Arbor, Mich.: Ann Arbor Science Pub. 1972.

the case. We now recognize seas and oceans as fragile environments and are able to measure the detrimental effect of our actions.

The water in the oceans is the most complicated chemical solution imaginable, and there is evidence that it has changed very little over millions of years. Because of this constancy, however, marine organisms have become highly specialized and intolerant to environmental change. Oceans are thus fragile ecosystems, quite susceptible to pollution.

A relief map of the ocean bottom reveals that there are two major areas of what we think of as ocean: the continental shelf and the deep oceans. The continental shelf, and especially the areas near major estuaries, are the most productive in terms of food supply. These also receive the greatest pollutional load. Many major estuaries have become so badly polluted that they have been closed to commercial fishing. Some large areas, such as the Baltic and Mediterranean Seas, are also in danger of becoming permanently damaged.

Although ocean disposal of wastewater is severely restricted in the United States, many major cities all over the world still discharge all of their untreated sewage into the oceans. This is usually done by pipelines running considerable distances from the shore and discharging through diffusers to achieve maximum dilution. The controversy continues as to the wisdom of using the oceans for wastewater disposal and what the long-term consequences might be.

CONCLUSION

Human activities naturally produce by-products that are often characterized as "waste." The more sophisticated the activity, generally the more dangerous the waste is and the more effort is needed to prevent the contamination of global water resources. In the next chapter, we discuss how engineers and scientists measure the pollutional potential of waste water and the quality of water.

PROBLEMS

2.1 The aerobic cycle, shown as Figure 2-3, is only for nitrogen, carbon, and sulfur. Phosphorus should also have been included in the cycle since it exists as organic phosphorus in living and dead tissue, decomposes to polyphosphates (such as $(P_2 O_7)^{-4}$, $(P_3 O_{10})^{-5}$, and so forth), and finally to the inorganic orthophosphate, PO_4^{-3}. Draw a phosphorus cycle similar to Figure 2-3.

2.2 Draw typical oxygen sag curves for the following wastes:

    a. potato waste (high BOD)
    b. electroplating waste (Cr, Co, and so forth)
    c. brick waste (clay)

Assume that the stream has a DO of 5 mg/L at the point of the pollution and a temperature of 20 °C.

2.3 Some researchers have suggested that the empirical analysis of some algae has the following chemical composition (by weight):

$$C_{106} H_{181} O_{45} N_{16} P$$

Suppose an analysis of a lake water yields the following:

$$C = 62 \text{ mg/L}$$
$$N = 1.0 \text{ mg/L}$$
$$P = 0.01 \text{ mg/L}$$

Which element would be limiting the growth of algae in this lake? (Show your calculations.)

2.4 A stream feeding a lake has an average flow of 1 cubic foot per second and a phosphate concentration of 10 mg/L. The water leaving the lake has a phosphate concentration of 5 mg/L.
    a. How much phosphate is deposited in the lake every year?
    b. Where does this phosphorus go? (The outflow is less than the inflow so it has to go somewhere.)
    c. Would you expect the average phosphate concentration to be higher near the surface or the bottom?
    d. Would you expect this lake to have accelerated eutrophication problems? Why?

2.5 Suppose you have a 55-gallon drum full of distilled water, and you add one frog weighing 0.2 pound. What is the concentration of frogs in terms of mg/L? Show all your calculations.

2.6 What could be done to maximize the silting in a reservoir?

2.7 The temperature soundings for a lake are as follows:

Depth (ft)	Water Temperature (°F)
surface	80
4	80
8	60
16	40
24	40
30	40

Draw a graph of depth versus temperature and label the hypolimnion, epilimnion, and thermocline.

2.8 If an industrial plant discharges solids at a rate of 5000 lb/day and if each person contributes 0.2 lb/day, what is the population equivalent of the waste?

2.9 Draw the DO sag curves you would expect in a stream from the following wastes. Assume the stream flow equals the flow of wastewater. (Do not calculate.)

Waste	Source	Demand for oxygen, BOD (mg/L)	Suspended Solids, SS (mg/L)	Phosphorus (mg/L)
A	Dairy	2000	100	40
B	Brick mfg.	5	100	10
C	Fertilizer mfg.	25	5	200
D	Plating plant	0	100	10

2.10 Starting with nitrogenous dead organic matter, follow N around the aerobic and anaerobic cycles by writing down all the various forms of nitrogen.

2.11 Why does eutrophication rarely occur in a stream?

2.12 Name six ways you can recognize an anaerobic stream (without instrumentation).

2.13 Suppose a stream with a velocity of 1 ft/sec, flow of 10 mgd and an ultimate BOD of 5 mg/L was hit with treated sewage at 5 mgd with an ultimate BOD (L_o) of 60 mg/L. The temperature of the stream water is 20 °C, at the point of the sewage discharge the stream is 90 percent saturated with oxygen and wastewater is at 30 °C and has no oxygen (see Table 3-1). Measurements show the deoxygenation constant $k_1' = 0.5$ and the reoxygenation $k_2' = 0.6$ both as days^{-1}. Calculate: (a) the oxygen deficit 1 mile downstream, (b) the minimum DO (the lowest part of the sag curve), and (c) the minimum DO (or maximum deficit) if the ultimate BOD of the treatment plant's effluent was 10 mg/L. Use a computer program if possible.

2.14 An industry discharges sufficient quantities of organic wastes to depress the oxygen sag curve to 2 mg/L 5 miles downstream, at which point the DO begins to increase. This DO is too low, and the state regulatory agency wants to fine them for noncompliance with standards. One solution is to build a wastewater treatment plant. This is expensive, and so the industry looks around for other means of resolving their problem.

A salesman tells the plant engineer about freeze-dried bacteria, which come to life in water and are especially adapted to the kind of organic waste the plant is discharging. The salesman suggests that the plant engineer dump the bacteria into the river at the point of the industrial waste discharge and that this will solve the problem.

Will it? Why or why not? Draw the DO sag curve before and after he adds the bacteria.

LIST OF SYMBOLS

D = deficit in dissolved oxygen, mg/L

D_a = initial dissolved oxygen deficit, mg/L

d = diversity index

k_1 = deoxygenation constant, $(\log_{10})$ $\sec^{-1}$

k_1' = deoxygenation constant, $(\log_e)$ $\sec^{-1}$

k_2 = reoxygenation constant, $(\log_{10})$ $\sec^{-1}$

k_2' = reoxygenation constant, $(\log_e)$ $\sec^{-1}$

L_o = ultimate biochemical oxygen demand, mg/L

m = rank assigned to low flows

n = number of individuals in all S species

n = number of years in low flow records

n_i = number of individuals in species i

T = temperature, °C

t = time, sec

t_c = critical time, time when the minimum DO occurs, sec

v = velocity, m/sec

y = oxygen used, mg/L

z = oxygen required for decomposition, mg/L

Appendix:
Mathematical Description
of the Dissolved
Oxygen Sag Curve

The effect of a certain type of waste on the oxygen level in a stream can be estimated quite accurately by calculating the rate of oxygen use, oxygen depletion, (Figure 2–A1). To make this calculation, we make three basic assumptions:

1. The amount of oxygen is *increased* by reaeration—O_2 dissolved into the water from the atmosphere.
2. The amount of oxygen is *decreased* as oxygen is used by microorganisms in the stream.
3. The competition between the increase and decrease results in an oxygen balance at any point in the stream.

The rate of oxygen depletion can be expressed as:

$$\frac{d[O_2]}{dt} = -k_1' z$$

where $[O_2]$ = concentration of DO
   z = the amount of oxygen still required at any time t, or the biochemical demand for oxygen remaining in the water, mg/L
   k_1' = deoxygenation constant (a function of the waste material, temperature, and so forth) in days^{-1}

The value of k_1' is measured in the laboratory, as is discussed in the next chapter.

The equation above is an example of the rate equation for a first-order chemical reaction. Integrating it yields

$$z = L_o e^{-k_1' t}$$

where L_o is the *ultimate oxygen demand,* or longterm need for oxygen, and z is the amount of oxygen still needed at any time t. Thus the amount of oxygen used at any time t must be

$$y = L_o - z$$

39

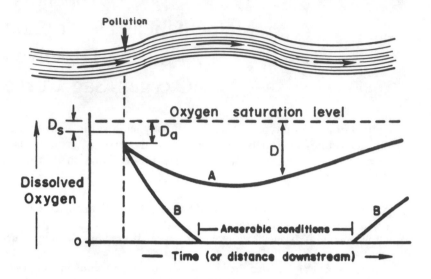

Figure 2–A1 Dissolved oxygen downstream from a source of organic pollution. Curve A depicts an oxygen sag without anaerobic conditions; curve B shows DO levels when pollution is sufficiently strong to create anaerobic conditions. D_a is the oxygen deficit in the stream after the stream has mixed with the pollutant.

This relationship is shown in Figure 2–A2. The term y is also called the *biochemical oxygen demand* (BOD). Substituting for z in the equation above, we have an expression for BOD:

$$y = BOD = L_o(1 - e^{-k'_1 t})$$

The biochemical oxygen *demand* is more correctly described as the dissolved oxygen *used*. In this text, we use the terms "oxygen demanded" and "oxygen used" synonymously.

The reoxygenation of a watercourse can be expressed as:

$$\text{Rate of reoxygenation} = k'_2 D$$

where D = deficit in dissolved oxygen or the difference between saturation (maximum dissolved oxygen the water can hold) and the actual DO, mg/L

k'_2 = reoxygenation constant, days^{-1}

The reoxygenation constant k'_2 may be estimated using generalized tables, such as Table 2–A1.

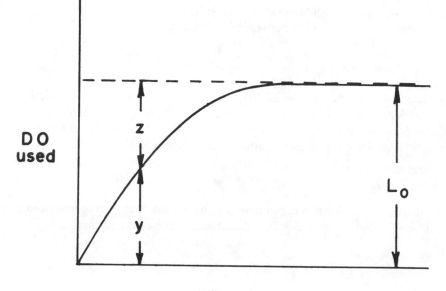

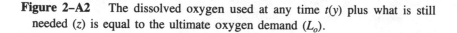

Time

Figure 2–A2 The dissolved oxygen used at any time $t(y)$ plus what is still needed (z) is equal to the ultimate oxygen demand (L_o).

If a stream is loaded with organic material, the simultaneous action of deoxygenation and reoxygenation forms the dissolved oxygen sag curve first discussed by Streeter and Phelps in 1925.

The shape of the oxygen sag curve, as shown in Figure 2–5, is the result of adding the rate of oxygen use and rate of supply. If the rate of use is great, as in the stretch of stream immediately after the pollution, the DO level drops because the supply rate cannot keep up with it. As the rate of oxygen use decreases (fewer high-energy readily decomposable organics) and the difference between oxygen saturation level and actual DO (the

Table 2–A1.   Reaeration Constants

	k'_2 at 20 °C[a], (days^{-1})
Small Ponds or Backwaters	0.1–0.23
Sluggish Streams	0.23–0.35
Large Streams, Low Velocity	0.35–0.46
Large Streams, Normal Velocity	0.46–0.69
Swift Streams	0.69–1.15
Rapids	> 1.15

[a]For temperatures other than 20 °C, $k'_{2_T} = k'_{2_{20}} = 1.024^{T-20}$.

deficit) is great, the supply begins to keep up with the use, and the DO will once again reach saturation levels.

This can be expressed mathematically as:

$$\frac{dD}{dt} = k_1'z - k_2'D$$

where all the terms are as defined above. The rate of change in the deficit (D) depends on the concentration of decomposable organic matter or the need by the microorganisms for oxygen (z) and the deficit at any time t. The need for oxygen at any time t was previously expressed as:

$$z = L_o e^{-k_1't}$$

where L_o = ultimate oxygen demand, or the maximum oxygen required in mg/L. Substituting this into the above expression and integrating,

$$D = \frac{k_1'L_o}{k_2' - k_1'} (e^{-k_1't} - e^{-k_2't}) + D_a e^{-k_2't}$$

where D_a = the initial oxygen deficit, at the point of pollution after the stream flow has mixed with the source of pollution, mg/L
         D = deficit at any time t, mg/L

Note that

$$D_a = \frac{D_s Q_s + D_p Q_p}{Q_s + Q_p}$$

where D_s = oxygen deficit in the stream directly upstream from the point of pollution, mg/L
         Q_s = stream flow, m³/s
         D_p = oxygen deficit in the pollutant stream, mg/L
         Q_p = flow rate of pollutant, m³/s.

Example 2.A1

Assume that a large stream has a reoxygenation constant k_2' of 0.4/day, a flow velocity of 5 miles per hour, and at the point of the pollution discharge the stream is saturated with oxygen at 10 mg/L. The wastewater flow rate is small compared to the stream flow, so the mixture is assumed to have a DO at saturation, and an oxygen demand of 20 mg/L. The deoxygenation constant is 0.2/day. What is the DO level 30 miles downstream?

Velocity = 5 mph, hence it takes 30/5 = 6 hr to travel 30 mi
         t = 6/24 = 1/4 days
         D_a = 0 since the stream is saturated

$$D = \frac{(0.2)(20)}{0.4 - 0.2} [e^{-0.2(0.25)} - e^{-0.4(0.25)}] = 20[e^{-0.05} - e^{-0.1}] = 1.0 \text{ mg/L.}$$

The DO is thus the saturation level minus the deficit, or $10 - 1.0 = 9.0$ mg/L.

Chapter 3

Measurement of Water Quality

Quantitative measurements of pollutants are obviously necessary before water pollution can be controlled. Measurement of these pollutants is, however, fraught with difficulties.

The first problem is that the specific materials responsible for the pollution are sometimes not known. The second difficulty is that these pollutants are generally at low concentrations, and very accurate methods of detection are therefore required.

Only a few of the many analytical tests available to measure water pollution are discussed in this chapter. A complete volume of analytical techniques used in water and wastewater engineering is compiled as *Standard Methods*.[1] This volume, now in its 16th edition, is the result of a need for standardizing test techniques. It is considered definitive in its field and has the weight of legal authority.

Many of the pollutants are measured in terms of milligrams of the substance per liter of water (mg/L). This is a weight/volume measurement. In many older publications pollutants are measured as parts per million (ppm), a weight/weight parameter.[2] If the liquid involved is water, these two units are identical since 1 milliliter of water weighs 1 gram. For pollutants present in very low concentrations (< 10mg/L), ppm is approximately equal to mg/L. Because of the possibility of some wastes not having the specific gravity of water, however, the ppm measure has been scrapped in favor of mg/L.

---

[1] *Standard Methods for the Examination of Water and Wastewater*, 16th ed. (WPCT, AWWA, APHA, 1988).

[2] In air pollution, however, ppm is in volume/volume.

A third commonly used parameter is percent, a weight/weight relationship. Obviously 10,000 ppm = 1 percent and this is equal to 10,000 mg/L only if 1 mL = 1 g.

SAMPLING

Some tests require the measurement to be conducted in the stream since the process of obtaining a sample may change the measurement. For example, if it is necessary to measure the dissolved oxygen in a stream, the measurement should be conducted right in the stream, or the sample must be extracted with great care to assure that no transfer of oxygen between the air and water (in or out) has occurred.

Most tests can be performed on a water sample taken from the stream. The process by which that sample is obtained, however, can greatly influence the result.

There are basically three types of samples:

1. grab
2. composite
3. flow weighed composite

The grab, as the name implies, simply measures a point. Its value is that it represents accurately the water quality at the moment of sampling, but obviously says nothing about the quality before or after the sampling.

The composite sample is obtained by taking a series of grab samples and mixing them together. The flow weighed composite is obtained by taking each sample so that the volume of the sample is proportional to the flow at that time. The last method is especially useful when daily loadings to wastewater treatment plants are calculated.

Whatever the technique or method, however, it is necessary to recognize that the analysis can only be as accurate as the sample, and often the sampling methodology is far more sloppy than the analytical determination.

DISSOLVED OXYGEN

Probably the most important measure of water quality is the dissolved oxygen. Oxygen, although poorly soluble in water, is fundamental to aquatic life. Without free dissolved oxygen, streams and lakes become uninhabitable to gill-breathing aquatic organisms. The amount of oxygen that can be dissolved in water is inversely proportional to temperature, and

Table 3-1.   Solubility of Oxygen

Temperature of Water (°C)	Saturation Concentration of Oxygen in Water (mg/L)
0	14.6
2	13.8
4	13.1
6	12.5
8	11.9
10	11.3
12	10.8
14	10.4
16	10.0
18	9.5
20	9.2
22	8.8
24	8.5
26	8.2
28	8.0
30	7.6

the maximum oxygen that can be dissolved in water at most ambient temperatures is about 9 mg/L. This saturation value decreases rapidly with increasing water temperature, as shown in Table 3-1. The balance between saturation and depletion is therefore tenuous.

The amount of oxygen dissolved in water is usually measured by an oxygen probe. The simplest (and historically the first) probe is shown in Figure 3-1. The principle of operation is that of a galvanic cell. If lead and silver electrodes are put in an electrolyte solution with a micro-ammeter between, the reaction at the lead electrode would be

$$Pb + 2OH^- \rightarrow PbO + H_2O + 2e^-$$

At the lead electrode, electrons are liberated that travel through the microammeter to the silver electrode where the following reaction takes place:

$$2e^- + \frac{1}{2}O_2 + H_2O \rightarrow 2OH^-$$

The reaction would not go unless free dissolved oxygen is available, and the microammeter would not register any current. The trick is to construct and calibrate a meter in such a manner that the electricity recorded is proportional to the concentration of oxygen in the electrolyte solution.

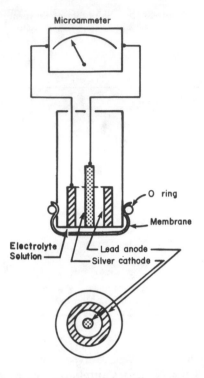

Figure 3-1   The galvanic cell oxygen probe.

In the commercial models the electrodes are insulated from each other with nonconducting plastic and are covered with a permeable membrane with a few drops of an electrolyte between the membrane and the electrodes. The amount of oxygen that travels through the membrane is proportional to the DO concentration. A high DO in the water creates a strong push to get through the membrane, while a low DO would force only limited O_2 through to participate in the reaction and thereby create electrical current. Thus the current registered is proportional to the oxygen level in solution.

BIOCHEMICAL OXYGEN DEMAND

Perhaps even more important than the determination of dissolved oxygen is the measurement of the rate at which this oxygen is used. A very low rate of use would indicate either clean water or that the available microorganisms are uninterested in consuming the available organics. A third possibility is that the microorganisms are dead or dying. (Nothing decreases oxygen consumption by aquatic microorganisms quite so well as a healthy slug of arsenic.)

The rate of oxygen use is commonly referred to as *biochemical oxygen demand* (BOD) introduced in Chapter 2. It is important to understand that BOD is not a measure of some specific pollutant. Rather it is a measure of the amount of oxygen required by bacteria and other microorganisms while stabilizing decomposable organic matter.

The BOD test was first used for measuring the oxygen consumption in a stream by filling two bottles with stream water, measuring the DO in one, and placing the other in the stream. In a few days the second bottle was retrieved and the DO measured. The difference in the oxygen levels was the BOD, or oxygen demand, in milligrams of oxygen used per liter of sample.

This test had the advantage of being very specific for the stream in question since the water in the bottle is subjected to the same environmental factors as is the water in the stream, and thus the result is an accurate measure of DO usage in that stream. It was impossible, however, to compare the results in different streams since three very important variables were not constant: temperature, time, and light.

Temperature has a pronounced effect on oxygen uptake (usage) with metabolic activity increasing significantly at higher temperatures. The time allotted for the test is also important since the amount of oxygen used increases with time. Light is also an important variable since most natural waters contain algae and oxygen can be replenished in the bottle if light is available. Different amounts of light would thus affect the final oxygen concentration.

The BOD test was finally standardized by requiring the test to be run in the dark at 20 °C for five days. This is defined as a five-day BOD, or BOD_5, or the oxygen used in the first five days. Although there appear to be some substantial scientific reasons why five days was chosen, it has been suggested that the possibility of preparing the samples on a Monday and taking them out on Friday, thus leaving the weekend free, was not the least important of these reasons.

The BOD test is almost universally run using a standard BOD bottle (about 300 mL volume) as shown in Figure 3–2. It is of course also possible to have a 2-day, 10-day, or any other day BOD. One measure used in some cases is *ultimate BOD* or the O_2 demand after a very long time.

If we measure the oxygen in several samples every day for five days, we may obtain curves such as Figure 3–3. Referring to this figure, sample A had an initial DO of 8 mg/L, and in five days this has dropped to 2 mg/L. The BOD_5 therefore is $8 - 2 = 6$ mg/L.

Sample B also had an initial DO of 8 mg/L, but the oxygen was used up so fast that it dropped to zero. If after five days we measure zero DO,

Figure 3-2   A BOD bottle.

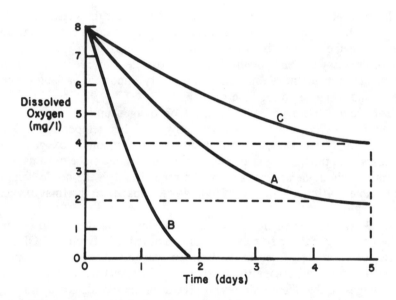

Figure 3-3   Typical oxygen uptake (use) curves in a BOD test.

we know that the BOD_5 of sample B was more than $8 - 0 = 8$ mg/L, but we do not know how much more since the organisms might have used more DO if it were available. For samples containing more than about 8 mg/L, dilution of the sample is therefore necessary.

Suppose sample C shown on the graph is really sample B diluted by 1:10. The BOD_5 of sample B is therefore

$$\frac{8 - 4}{0.1} = 40 \text{ mg/L}$$

It is possible to measure the BOD of any organic material (e.g., sugar) and thus estimate its influence on a stream, even though the material in its original state might not contain the necessary organisms. *Seeding* is a process in which the microorganisms that create the oxygen intake are added to the BOD bottle.

Suppose we used the water previously described by the A curve as seed water since it obviously contains microorganisms (it has a 5-day BOD of 6 mg/L). We now put 100 mL of an unknown solution into a bottle and add 200 mL of seed water, thus filling the 300-mL bottle. Assuming that the initial DO of this mixture was 8 mg/L and the final DO was 1 mg/L, the total oxygen used was 7 mg/L. But some of this was due to the seed water, since it also has a BOD, and only a portion was due to the decomposition of the unknown material. The DO uptake due to the seed water was

$$6 \times \left(\frac{2}{3}\right) = 4 \text{ mg/L}$$

since only ⅔ of the bottle was the seed water, which has a BOD of 6 mg/L. The remaining oxygen uptake (7 − 4 = 3 mg/L) must have been due to the unknown material.

If the seeding and dilution methods are combined, the following general formula is used to calculate the BOD:

$$BOD(\text{mg/L}) = \frac{(I - F) - (I' - F')(X/Y)}{D}$$

where
I = initial DO of bottle with sample and seeded dilution water
F = final DO of bottle with sample and seeded dilution water
I' = initial DO of bottle with seeded dilution water
F' = final DO of bottle with seeded dilution water
X = mL seeded dilution water in sample bottle
Y = mL in bottle with only seeded dilution water
D = dilution of sample

Example 3.1

Calculate the BOD_5, given the following data:
Temperature of sample, 20 °C
Initial DO is saturation
Dilution is 1:30 with seeded dilution water
Final DO of seeded dilution water is 8 mg/L
Final DO bottle with sample and seeded dilution water is 2 mg/L
Volume of BOD bottle is 300 mL.

From Table 3-1, at 20 °C saturation is 9.2 mg/L; hence this is the initial DO. Since the BOD bottle contains 300 mL, a 1:30 solution would have

10 mL of sample and 290 mL of seeded dilution water, so that D = 10/300.

$$BOD_5 = \frac{(9.2 - 2) - (9.2 - 8)(290/300)}{0.033} = 183 \text{ mg/L}$$

If, instead of stopping the test after five days, we allowed the reactions to proceed and measured the DO each day, we might get a curve like Figure 3–4. Note that some time after five days the curve takes a sudden jump. This is due to the exertion of oxygen demand by the microorganisms that decompose nitrogenous organics and are converting these to the stable nitrate, NO_3^-. The curve is thus divided into nitrogenous and carbonaceous BOD areas. Note also the definition of the ultimate BOD.[3]

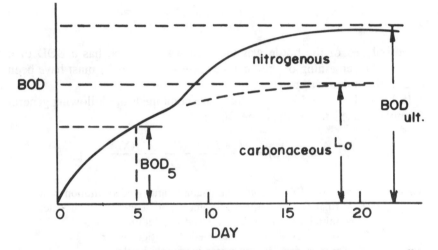

Figure 3–4   Long-term BOD. Note that BOD_{ult} includes nitrogenous as well as the ultimate carbonaceous BOD (L_o).

It is important to remember that BOD is a measure of oxygen use or potential use. An effluent with a high BOD can be harmful to a stream if the oxygen consumption is great enough to cause anaerobic conditions. Obviously, a small trickle going into a great river will have negligible effect, regardless of the mg/L of BOD involved. Similarly, a large flow into a small stream can seriously affect the stream even though the BOD might be low. Accordingly, engineers often talk of "pounds of BOD,"

---

[3] This, however, is different from the definition of ultimate BOD given earlier in this chapter. The latter would more accurately be termed *ultimate carbonaceous BOD*.

a value calculated by multiplying the concentration by the flow rate, with a conversion factor, so that

$$\text{lb BOD/day} = [\text{mg/L BOD}] \times \left[\begin{array}{c} \text{flow in million} \\ \text{gallons per day} \end{array}\right] \times 8.34$$

The BOD of most domestic sewage is about 200 mg/L. Many industrial wastes run as high as 30,000 mg/L. The potential detrimental effect of an untreated dairy waste, which might have a BOD of 20,000 mg/L, is quite obvious.

As discussed in Chapter 2, the carbonaceous part of the BOD curve can be modeled using the equation

$$y = L_o(1 - e^{-k_1' t})$$

where
y = BOD_t, or the amount of oxygen demanded by the micro-organisms at some time t, mg/L
L_o = the ultimate carbonaceous demand for oxygen, mg/L
k_1' = rate constant previously termed the deoxygenation constant, days^{-1}
t = time, days.

It should be again emphasized that this equation is applicable only to the carbonaceous BOD curve, and L_o is defined as the ultimate carbonaceous BOD.

For streams and rivers with travel times greater than about five days, the ultimate demand for oxygen must include the nitrogenous demand. Although the use of BOD_{ult} (carbonaceous plus nitrogenous) in dissolved oxygen sag calculations is not strictly accurate, it is often assumed that the ultimate BOD can be calculated as

$$\text{BOD}_{\text{ult}} = a(\text{BOD}_5) + b(\text{KN})$$

where
KN = Kjeldahl nitrogen (organic plus ammonia mg/L)
a and b = constants.

The state of North Carolina, for example, uses $a = 1.2$ and $b = 4.0$ for calculating the ultimate BOD, which is then substituted for L_o in the dissolved oxygen sag equation.

CHEMICAL OXYGEN DEMAND

Among the many drawbacks of the BOD test the most important is that it takes five days to run. If the organics were oxidized chemically instead of biologically, the test could be shortened considerably. Such oxidation is accomplished with the *chemical oxygen demand* (COD) test. Because nearly all organics are oxidized in the COD test and only some are decomposed during the BOD test, COD values are always higher than BOD values. One example of this is wood pulping wastes where compounds such as cellulose are easily oxidized chemically (high COD) but are very slow to decompose biologically (low BOD).

Potassium dichromate is generally used as an oxidizing agent. It is an inexpensive compound that is available in very pure form. A known amount of this compound is added to a measured amount of sample and the mixture boiled. The reaction, in unbalanced form, is

$$C_X H_Y O_Z + Cr_2 O_7^= + H^+ \xrightarrow{\Delta} CO_2 + H_2O + Cr^{+++}$$

(organic) (dichromate)

After boiling with an acid, the excess dichromate (not used for oxidizing) is measured by adding a reducing agent, usually ferrous ammonium sulfate. The difference between the chromate originally added and the chromate remaining is the chromate used for oxidizing the organics. The more chromate used, the more organics were in the sample, and hence the higher the COD.

TOTAL ORGANIC CARBON

Since the ultimate oxidation of organic carbon is to CO_2, the total combustion of a sample will yield some significant information on the amount of organic carbon present in a wastewater sample. Without elaboration, this is done by allowing a little bit of the sample to be burned in a combustion tube and measuring the amount of CO_2 emitted. This test is not widely used at present, mainly because of the expensive instrumentation required. It has significant advantages over the BOD and COD tests, however, and will undoubtedly be more widely used in the future.

TURBIDITY

Water that is not clear but is "dirty," in the sense that light transmission is inhibited, is known as *turbid water*. Turbidity can be caused by many materials, some of which are discussed in Chapter 2. In the treatment of water for drinking purposes, turbidity is of great importance first because of the aesthetic considerations and second because pathogenic organisms can hide on (or in) the tiny colloidal particles.

The standard method of measuring turbidity until recently was with the Jackson Candle Turbidimeter, first developed in 1900. It consists of a long flat-bottomed glass tube under which a candle is placed. The turbid water is poured into the glass tube until the outline of the flame is no longer visible. Obviously, with very turbid water only a little water is required to obscure the flame. The centimeters of water in the tube are then measured and compared to the standard turbidity unit:

$$1 \text{ mg/L } SiO_2 = 1 \text{ unit of turbidity}$$

Recent years have seen the development of nephelometers, photometers that measure the intensity of scattered light. The opaque particles that cause turbidity scatter light so that the scattered light measured at right angles to a beam of incident light is proportional to the turbidity. These instruments have completely replaced the antiquated Jackson Candle. The standard unit of turbidity is now the Nephelometric Turbidity Unit (NTU) instead of the Jackson Turbidity Unit (JTU).

COLOR AND ODOR

Color and odor are both important measurements in water treatment. Along with turbidity they are called the *physical* parameters of drinking water quality. Color and odor are important from the standpoint of aesthetics. If water looks colored or smells bad, people will instinctively avoid using it, even though it might be perfectly safe from the public health aspect. Both bad color and odor can be, and often are, caused by organic substances such as algae or humic compounds.

Color is measured by comparison with standards. Colored water made with potassium chloroplatinate when tinted with cobalt chloride closely resembles the color of many natural waters. Where multicolored industrial wastes are involved, such color measurement is meaningless.

Odor is measured by successive dilutions of the sample with odor-free water until the odor is no longer detectable. This test is obviously subjective and depends entirely on the olfactory senses of the tester.

pH

The pH of a solution is a measure of the hydrogen ion concentration. An acidic solution has an excess of hydrogen ions, a basic or alkaline solution has a dearth of H^+ (and an excess of OH^-) ions.

Pure water, HOH, dissociates very slightly. On dissociation, of course, water yields an equal number of H^+ and OH^- ions. The degree to which pure water dissociates is indicated by the dissociation constant K_w:

$$K_w = [H^+][OH^-] = 10^{-14}$$

where $[H^+]$ = the concentration of H^+ ions
$[OH^-]$ = the concentration of OH^- ions

In pure water

$$[H^+] = [OH^-] = 10^{-7} \text{ moles/liter}$$

The product of the concentrations of hydrogen and hydroxyl ions in water solution is *always* 10^{-14}. Thus if the concentration of H^+ is 10^{-5} moles/liter, the hydroxyl ion concentration is

$$[OH^-] = \frac{10^{-14}}{10^{-5}} = 10^{-9} \text{ moles/liter}$$

Such a solution is acidic since there is an excess of hydrogen ions over 10^{-7} moles/liter.

The hydrogen ion concentration is so important in aqueous solutions that an easier method of expressing it has been devised. Instead of speaking in terms of moles per liter, we can take the negative logarithm of $[H^+]$ so that

$$pH = -\log[H^+] = \log\frac{L}{[H^+]}$$

The pH value is the negative power to which 10 must be raised to equal the hydrogen ion concentration or

$$[H^+] = 10^{-pH}$$

For a neutral solution $[H^+]$ is 10^{-7}, or pH = 7. Obviously for greater hydrogen ion concentrations the pH is lower, and for greater hydroxide ion concentrations (or lower hydrogen ion concentrations since the products, as you recall, are constant) the pH rises. The pH range of dilute solutions is from 0 (very acidic, 1 mole of H^+ ions per liter) to 14 (very alkaline).

The measurement of pH is now almost universally by electronic means. Electrodes sensitive to hydrogen ion concentration (strictly speaking, the hydrogen ion activity) convert the signal to electrical current.

pH is important in almost all phases of water and wastewater treatment. Aquatic organisms are sensitive to pH changes, and biological treatment requires either pH control or monitoring. In water treatment pH is important in ensuring proper chemical treatment as well as in disinfection and corrosion control. Mine drainage often involves the formation of sulfuric acid (high H^+ concentration) that is extremely detrimental to aquatic life. Continuous acid deposition from the atmosphere, commonly called "acid rain" (see Chapter 17), can lower the pH of a lake substantially.

SOLIDS

One of the main problems with wastewater treatment is that much of the contents of the wastewater is actually solid. The separation of these solids from the water is in fact one of the primary objectives of treatment.

Strictly speaking, in wastewater anything other than water would be classified as a solid. The usual definition of solids, however, is the residue on evaporation at 103 °C (slightly higher than the boiling point of water). The solids thus measured are known as *total solids*. Total solids can be divided into two fractions: *dissolved* and *suspended solids*. If we put a teaspoonful of common table salt in a glass of water, the salt will dissolve. The water will not look any different, but the salt will remain behind if we evaporate the water. A spoonful of sand, however, will not dissolve and will remain as sand grains in the water. The salt is an example of dissolved solids while the sand would be measured as suspended solids.

The separation of suspended solids from dissolved solids is by means of a special crucible, called a *Gooch crucible*. As shown in Figure 3–5, the Gooch crucible has holes on the bottom on which a fiberglass filter is placed. The sample is then drawn through the crucible with the aid of a vacuum. The suspended material is retained on the filter while the dis-

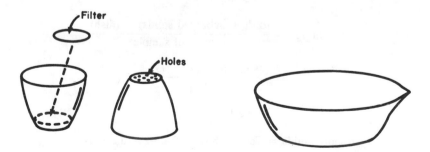

Figure 3–5 The Gooch crucible for suspended solids (with filter) and the evaporating dish for total solids.

solved fraction passes through. If the initial dry weight of the crucible and filter are known, the subtraction of this from the total weight of crucible, filter, and the dried solids caught on the filter yields the weight of suspended solids, expressed as mg/L.

Solids can be classified in another way: those that are volatilized at a high temperature and those that are not. The former are known as *volatile solids,* the latter as *fixed solids.* Usually volatile solids are organic. Obviously, at 600 °C, the temperature at which the combustion takes place, some of the inorganics are decomposed and volatilized, but this is not considered a serious drawback.

The relationship between the total solids and the total volatile solids can best be illustrated by an example.

Example 3.2

Given the following data:

- weight of dish (such as shown in Figure 3–6) = 48.6212 g
- 100 mL of sample is placed in dish and evaporated. Weight of dish and dry solids = 48.6432 g
- Dish is placed in 600 °C furnace, then cooled. Weight = 48.6300 g

Find the total, the volatile and fixed solids.

$$\text{Total Solids} = \frac{(\text{dish} + \text{dry solids}) - (\text{dish})}{\text{volume of sample}}$$

$$= \frac{(48.6432) - (48.6212)}{100}$$

$$= 220 \times 10^{-6} \text{g/mL}$$

$$= 220 \times 10^{-3} \text{mg/mL}$$

$$= 220 \text{ mg/L}$$

$$\text{Fixed Solids} = \frac{(\text{dish} + \text{unburned solids}) - (\text{dish})}{\text{volume of sample}}$$

$$= \frac{(48.6300) - (48.6212)}{100}$$

$$= 88 \text{ mg/L}$$

$$\text{Volatile Solids} = \text{Total Solids} - \text{Total Fixed Solids}$$

$$= 220 - 88 = 132 \text{ mg/L}$$

It is often necessary to measure the volatile fraction of suspended material since this is a quick (if gross) measure of the amount of microorganisms present. The *volatile suspended solids* are determined by simply burning the material in the Gooch crucible and weighing again. The loss in weight is interpreted as volatile suspended solids.

NITROGEN

Recall from Chapter 2 that nitrogen is an important element in biological reactions. Nitrogen can be tied up in high-energy compounds such as amino acids and amines. In this form the nitrogen is known as organic nitrogen. One of the intermediate compounds formed during biological metabolism is ammonia. Together with organic nitrogen, ammonia is considered an indicator of recent pollution. These two forms of nitrogen are often combined in one measure, known as *Kjeldahl nitrogen* after the scientist who first suggested the analytical procedure.

Aerobic decomposition eventually leads to the nitrite (NO_2^- and finally nitrate NO_3^-) nitrogen forms. A high-nitrate and low-ammonia nitrogen therefore suggests that pollution has occurred but quite some time ago.

All of the above forms of nitrogen can be measured analytically by colorimetric techniques. The basic idea of colorimetry is that the ion in question will combine with some compound and form a color and that the intensity of this color is proportional to the original concentration of the ion. For example, ammonia can be measured by adding a compound called *Nessler reagent* to the unknown sample. This reagent is a solution of potassium mercuric iodide, K_2HgI_4, and combines with ammonium ions to form a yellow-brown colloid. Since Nessler reagent is in excess, the amount of colloid formed is proportional to the concentration of ammonium ions in the sample.

The color is measured either by visual comparison (in long tubes called Nessler tubes) or photometrically. The basic workings of a photometer, illustrated in Figure 3–6, consist of a light source, a filter, the sample, and a photo cell. The filter allows only certain wavelengths of light to pass through thus lessening interferences and increasing the sensitivity of the photocell, which converts light energy to electrical current. An intense color will allow only a limited amount of light to pass through and thus create little current. On the other hand, a sample containing very little of the chemical in question will be clear, allow almost all of the light to pass through, and set up a substantial current. If the color intensity (and hence light absorbance) is directly proportional to the concentration of the unknown ion, the color formed is said to obey Beer's law:

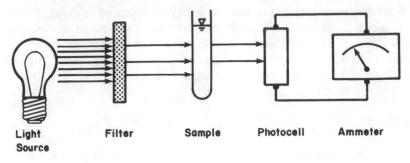

Light   Filter   Sample   Photocell   Ammeter
Source

Figure 3–6   Elements of a filter photometer.

$$I/I_o = e^{-acx}$$

or

$$\ln I_o/I = acx$$

where I = the intensity of light after it has passed through the sample
I_o = the intensity of light incident on the sample
a = the "absorption coefficient" of the colored compound
x = the path length of light through the sample
c = the concentration

A photometer can be used to measure ammonia concentration by measuring the absorbance of light by samples containing known ammonia concentration and comparing the absorbance of the unknown sample to these standards.

Example 3.3
Several known samples and an unknown sample were treated with Nessler reagent and the color measured with a photometer. Find the ammonia concentration of the unknown sample.

Sample	% Absorbance
Standards: 0 mg/L ammonia (Distilled water)	0
1 mg/L ammonia	6
2 mg/L ammonia	12
3 mg/L ammonia	18
4 mg/L ammonia	24
Unknown sample	15

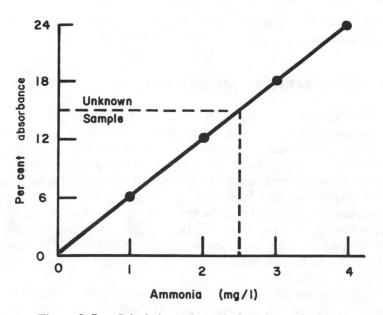

Figure 3-7   Calculation using colorimetric standards.

A plot (Figure 3-7) of ammonia concentration of the standards versus percent absorbance results in a straight line (Beer's Law is adhered to). We can then enter this chart at 15 percent absorbance (the unknown) and read the concentration of ammonia in our unknown as 2.5 mg/L.

PHOSPHATES

The importance of phosphorous compounds in the aquatic environment is discussed in Chapter 2. Phosphorus in wastewater can be either inorganic or organic. Although the greatest single source of inorganic phosphorus is synthetic detergents, organic phosphorus is found in food and human waste as well. All phosphates in nature will, by biological action, eventually revert to inorganic forms to be again used by the plants in making high-energy material.

Ever since phosphates were indicted as one culprit in lake eutrophication, the measurement of total phosphate has assumed considerable importance. Total phosphates can be measured by first boiling the sample in acid solution, which converts all the phosphates to the inorganic forms. From that point the test is colorimetric, using a chemical that, when combined with phosphates, produces a color directly proportional to the phosphate concentration.

BACTERIOLOGICAL MEASUREMENTS

From the public health standpoint the bacteriological quality of water is as important as the chemical quality. A number of diseases can be transmitted by water among them typhoid and cholera. It is one thing, however, to declare that water must not be contaminated by pathogens (disease-causing organisms) and another to determine the existence of these organisms. First, there are many pathogens. Each has a specific detection procedure and must be screened individually. Second, the concentration of these organisms can be so small as to make their detection impossible. It is a perfect example of the proverbial needle in a haystack. And yet only one or two organisms in the water might be sufficient to cause an infection.

How then can we measure for bacteriological quality? The answer lies in the concept of indicator organisms. The indicator most often used is a group of microbes of the family *Escherichia coli* or *E. coli* (often called coliform bacteria), which are organisms normal to the digestive tracts of warm-blooded animals. In addition to that attribute, coliforms are:

- plentiful, hence not difficult to find
- easily detected with a simple test
- generally harmless except in unusual circumstances
- hardy, surviving longer than most known pathogens

Coliforms have thus become universal indicator organisms. But the presence of coliforms does not prove the presence of pathogens. If a large number of coliforms are present, there is a good chance of recent pollution by wastes from warm-blooded animals, and therefore the water *may* contain pathogenic organisms.

This last point should be emphasized. The presence of coliforms does not mean that there are pathogens in the water. It simply means that there *might* be. A high coliform count is thus suspicious and the water should not be consumed (although it may be perfectly safe).

The simplest way to measure for coliforms is to filter a sample through a sterile filter thus capturing any coliforms. The filter is then placed in a petri dish containing a sterile agar that soaks into the filter and promotes the growth of coliforms while inhibiting other organisms. After 24 to 48 hours of incubation the number of shiny black dots, indicating coliform colonies, is counted. If we know how many milliliters were poured through the filter, the concentration of coliforms can be expressed as coliforms/mL.

VIRUSES

Because of their minute size and extremely low concentrations and the need to culture them on living tissues, pathogenic (or animal) viruses are fiendishly difficult to measure. Because of this problem, there are as yet no standards for viral quality of water supplies as there are for pathogenic bacteria.

One possible method of overcoming this difficulty is to use an indicator organism much like the coliform group is used as an indicator for bacterial contamination. This can be done by using a bacteriophage, a virus that attacks only a certain type of bacterium. For example, coliphages attack coliform organisms and because of their association with wastes from warm-blooded animals seem to be an ideal indicator. The test for coliphage is performed by inoculating a petri dish containing an ample supply of a specific type of coliform with the wastewater sample. Coliphages will attack the coliforms, leaving visible spots or plaques that can be counted so that an estimate can be made of the number of coliphages per known volume.

HEAVY METALS

The increasing problem of heavy metals in industrial effluent was mentioned in Chapter 2. Heavy metals such as arsenic and mercury can harm fish even at very low concentrations. Consequently, the method of measuring these ions in water must be very sensitive.

The method of choice for measuring heavy metals is *atomic absorption spectrophotometry*. In this method, a solution of lanthanum chloride is added to the sample, and the treated sample is sprayed into the flame with an atomizer. Each metallic element imparts a characteristic color to the flame whose intensity is then measured spectrophotometrically.

TRACE TOXIC ORGANICS

Very low concentrations of chlorinated hydrocarbons and other agrichemical residues in water can be assayed by *gas chromatography*. Oil residues in water are generally measured by extracting the water sample with freon and then evaporating the freon moiety and weighing the residue from this evaporation.

CONCLUSION

Discussed above are only some of the most important tests used in water pollution control. *Standard Methods,* for example, contains over 500 analytical procedures, many of which can be performed only with special equipment and skilled technicians.

Understanding this and realizing the complexity, variation, and objectives of some of the measurements of water pollution, how would you answer someone who brings a jug of water to your office, sets it on your desk, and asks, "Can you tell me how much pollution is in this water?"

PROBLEMS

3.1 Given the following BOD_5 test results:
  Initial DO   8 mg/L
  Final DO   0 mg/L
  Dilution   1/10
what can you say about
  a. BOD_5?
  b. BOD ultimate?
  c. COD?

3.2 If you had two bottles full of lake water and kept one in the dark and the other in daylight, which one would have a higher DO after a few days? Why?

3.3 Name three substances you would need to seed if you wanted to measure their BOD.

3.4 The following data were obtained for a sample:
  total solids = 4000 mg/L
  suspended solids = 5000 mg/L
  volatile suspended solids = 2000 mg/L
  fixed suspended solids = 1000 mg/L.
Which of these numbers is questionable and why?

3.5 A water has a BOD_5 of 10 mg/L. The initial DO in the BOD bottle was 8 mg/L and the dilution was 1 to 10. What was the final DO in the BOD bottle?

3.6 If the BOD_5 of a waste is 100 mg/L, draw a curve showing the effect of adding progressively higher doses of chromium (a toxic chemical) on the BOD_5.

3.7 Consider the following data from a BOD test:

Day	DO (mg/L)	Day	DO (mg/L)
0	9	5	6
1	9	6	6
2	9	7	4
3	8	8	3
4	7	9	3

What is the (a) BOD_5, (b) ultimate carbonaceous BOD, and (c) ultimate BOD? Why was there no oxygen used until the third day? If the sample were "seeded," would the final DO have been higher or lower and why?

3.8 An industry discharges 10 million gallons a day of a waste that has a BOD_5 of 2000 mg/L. How many pounds of BOD_5 are discharged?

3.9 Given the same standard ammonia samples as in Example 3.3, if your unknown measured 20 percent absorbance, what is the ammonia concentration in the unknown?

3.10 If coliform bacteria are to be used as an indicator of viral pollution as well as an indicator of bacterial pollution, what attributes must the coliform organisms have (relative to viruses)?

3.11 Draw a typical BOD curve. Label the (a) carbonaceous BOD, (b) nitrogenous BOD, (c) ultimate BOD, and (d) 5-day BOD. On the same graph, plot the BOD curve if the test had been run at 30°C instead of at the usual temperature. Also plot the BOD curve if a substantial amount of toxic materials were added to the sample.

LIST OF SYMBOLS

BOD = biochemical oxygen demand, mg/L
BOD_5 = five-day BOD
  D = dilution (volume of sample/total volume)
  F = final BOD of sample, mg/L
  F' = final BOD of seeded dilution water, mg/L
  I = initial BOD of sample, mg/L
  I' = initial BOD of seeded dilution water, mg/L
  X = seeded dilution water in sample bottle, mL
  Y = volume of BOD bottle, mL

Chapter 4

Water Supply

A supply of water is critical to the survival of life. People, animals, and plants all need water to drink. In addition, basic functions of a society require water—cleaning for public health, consumption for industrial processes, and cooling for electricity generation. In this chapter, we discuss the supply of water in terms of the hydrologic cycle and water availability, groundwater supplies, surface water supplies, and water transmission. The direction of our discussion is that sufficient water supplies exist for the world and the nation as a whole, but many areas are water poor while others are water rich. The trick in water supply is to engineer the supply and transmission of clean water from one area to another. In Chapter 5, we address treatment methods available to clean up the water once it reaches the areas of demand.

THE HYDROLOGIC CYCLE AND WATER AVAILABILITY

The concept of the hydrologic cycle is a useful starting point to begin our study of water supply. Illustrated in Figure 4-1, this cycle is the precipitation of water from clouds and infiltration into the ground or runoff into surface watercourses followed by evaporation and transpiration of the water back into the atmosphere.

The rates of precipitation and evaporation/transpiration help define the baseline quantity of water available for human consumption. *Precipitation* is the term applied to all forms of moisture originating in the clouds and falling to the ground. A range of instruments and technologies has been developed for measuring the amount and intensity of rain, snow, sleet, and hail. The average depth of precipitation over a given region, either on a storm, seasonal, or annual basis, is required in many water availability studies. Any open receptacle with vertical sides has application as a rain

67

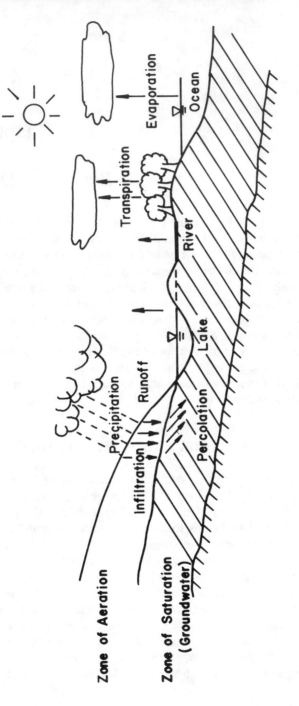

Figure 4-1   The hydrologic cycle.

gauge, but varying wind and splash effects must be considered if amounts collected by different gauges are to be compared.

Evaporation and *transpiration* help define the baseline quantity of water that is available. The same meteorological factors that influence evaporation are at work in the transpiration process: Solar radiation, ambient air temperature, humidity, and wind speed, as well as the amount of soil moisture available to the plants, impact the rate of transpiration. Many measurements for transpiration are made with a phytometer, a large vessel filled with soil and potted with selected plants. The surface of the soil is hermetically sealed to prevent evaporation; thus the only escape of moisture is through transpiration, which can be determined by weighing the plant-vessel system at time intervals up to the life of the plant. Because it is impossible to simulate natural conditions, the results of phytometer tests are of limited value. However, they can be used as an index of water demand by a crop under field conditions, and thus relate to the calculations that help an engineer determine the water supply requirements for that crop. Generally speaking, the terms evaporation and transpiration are linked by engineers into *evapotranspiration,* or the total water loss to the atmosphere by the two processes.

GROUNDWATER SUPPLIES

Groundwater is both an important direct source of supply tapped by wells and a significant indirect source of supply since a large portion of the flow to streams is derived from subsurface water.

Near the surface of the earth in the *zone of aeration* soil pore spaces contain both air and water. This zone, which may have a zero thickness in swamplands and be several hundred feet thick in mountainous regions, contains three types of moisture. *Gravity water* is in transit after a storm through the larger soil pore spaces. *Capillary water* is drawn through small pore spaces by capillary action, and is available for plant uptake. *Hygroscopic moisture* is water held in place by molecular forces during all except the driest climatic conditions. Moisture from the zone of aeration cannot be tapped as a water supply source.

On the other hand, the *zone of saturation* offers water in a quantity that is directly available. In this zone, located below the zone of aeration, the pores are filled with water, and this is what we consider *groundwater.* The stratum that contains a substantial amount of groundwater is called an *aquifer.* At the surface between the two zones, labeled the *water table* or *phreatic surface,* the hydrostatic pressure in the groundwater is equal to atmospheric pressure. An aquifer may extend to great depths, but because the weight of overburden material generally closes pore spaces, little water is found at depths greater than 600 m (2000 ft). The water readily available from an aquifer is that which will drain by gravity. Each soil type thus has a *specific yield,* defined as the volume of water, expressed as a percent of the total volume of water in the aquifer that will drain freely from the aquifer.

Besides offering a source of well water, aquifers also combine with precipitation to feed surface watercourses. Streams, rivers, and lakes also offer options for water supply, and they are addressed next.

SURFACE WATER SUPPLIES

Surface water supplies are not as reliable as groundwater sources since quantities often fluctuate widely during the course of a year, or even a week, and quality is restricted by various sources of pollution. If a river has an average flow of 10 cubic feet per second (cfs), it means often the flow is less than 10 cfs. If a community wishes to use this for a water supply, its demands for the water should be considerably less than 10 cfs in order to be assured of a reasonably dependable supply.

The variation in the river flow can be so great that even a small demand cannot be met during dry periods in many parts of the nation, and storage facilities must be constructed to hold the water during wet periods so it can be saved for the dry ones. The objective is to build these reservoirs sufficiently large to have dependable supplies.

One method of arriving at the proper reservoir size is by using the *mass curve.* Basically, the total flow in a stream at the point of a proposed reservoir is summed and plotted against the time. On the same curve the water demand is plotted and the difference between the total water flowing in and the water demanded is the quantity that the reservoir must hold if the demand is to be met. The method is illustrated by the following example problem:

Example 4.1

A reservoir is needed to provide a constant flow of 15 cfs. The monthly stream flow records, in total cubic feet, are:

Month	J	F	M	A	M	J	J	A	S	O	N	D
Cubic feet of water ($\times 10^6$)	50	60	70	40	32	20	50	80	10	50	60	80

We can calculate the storage requirement by plotting the cumulative water as in Figure 4–2. Note that for January we plot 50 million cubic feet, for February we add 60 to that and plot 110 million cubic feet, and so on.

The demand for water is constant at 15 cfs, or

$$15 \; \frac{\text{cubic feet}}{\text{sec}} \times 60 \; \frac{\text{sec}}{\text{min}} \times 60 \; \frac{\text{min}}{\text{hr}} \times 24 \; \frac{\text{hr}}{\text{day}} \times 30 \; \frac{\text{days}}{\text{month}}$$

$$= 38.8 \times 10^6 \; \text{cubic feet/month}$$

This can be represented as a sloped line in Figure 4–2, and plotted on the curved supply line. Note that the stream flow in May was lower than the demand, and this was the start of a drought lasting into June. The demand slope was greater than the supply, and thus the reservoir had to make up the deficit. In July the rains came and the supply increased until the reservoir could be filled up again late in August. The reservoir capacity needed to get through that particular drought was 60×10^6 cubic feet. A second

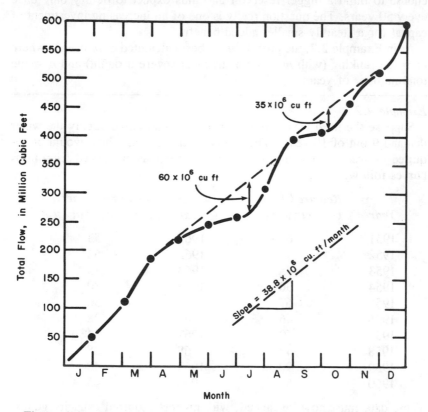

Figure 4–2   Mass curve for determining required reservoir capacity.

drought, starting in September, lasted into November and required 35×10^6 cubic feet of capacity. If the municipality therefore had a reservoir with at least 60×10^6 cubic feet capacity, they could have drawn water from it throughout the year.

A mass curve such as Figure 4–2 is actually of little use if only limited stream flow data are available. One year's data yield very little information about long-term variations. For example, was the drought in the above example that required 60 million cubic feet of storage the worst drought in 20 years, or was the year shown actually a fairly wet year?

To get around this problem, it is necessary to predict statistically the recurrence of events such as droughts and then to design the structures according to a known risk. Water supplies are often designed to meet demands 19 out of 20 years. In other words, once in 20 years the drought will be so severe that the reservoir capacity will not be adequate. If running out of water once every 20 years is not acceptable, the people can choose to build a bigger reservoir and thus expect to be dry only once every 50 years. The question really is one of an increasing investment of capital for a steadily smaller added benefit.

As in Example 2.2, the probability is best calculated as $m/n + 1$, where m = the ranking (with $m = 1$ for the most severe drought) and n = the total number of years.

Example 4.2

Suppose the need was to build a reservoir that could supply the water demand 9 out of 10 years. The reservoir capacities, which would be required in order to prevent running out of water as calculated from mass curves follow:

Year	Required Reservoir Capacity, mil cu m	Year	Required Reservoir Capacity, mil cu m
1951	60	1961	53
1952	40	1962	62
1953	85	1963	73
1954	30	1964	80
1955	67	1965	50
1956	46	1966	38
1957	60	1967	34
1958	42	1968	28
1959	90	1969	40
1960	51	1970	45

These data must now be ranked, with highest required capacity getting rank 1, the next highest 2, and so on ($n = 20$).

Rank	Capacity	$\dfrac{m}{n+1}$	Rank	Capacity	$\dfrac{m}{n+1}$
1	90	0.05	11	51	0.52
2	85	0.10	12	46	0.57
3	80	0.14	13	45	0.61
4	73	0.19	14	42	0.66
5	67	0.23	15	40	0.71
6	62	0.28	16	40	0.76
7	60	0.33	17	38	0.81
8	60	0.38	18	34	0.85
9	55	0.43	19	30	0.90
10	53	0.48	20	28	0.95

A semilog plot often yields an acceptable straight line. The 1 year in 10 drought has a probability of $m/n + 1 = 2/20 + 1 = 0.10$ of occurring. Using the curve in Figure 4–3, we can pick the 10 percent probability drought as 82 million cubic meters.

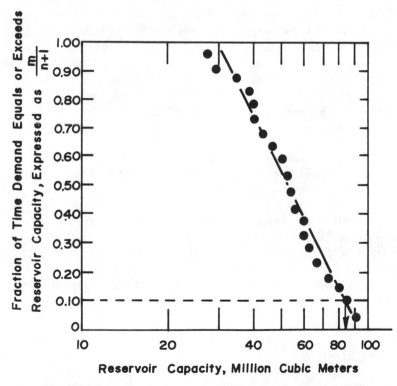

Figure 4–3 Frequency analysis of reservoir capacity.

This procedure is known as a *frequency analysis* of recurring natural events such as droughts. In the above example we selected a "10-year drought" or the drought that on the average occurs once every 10 years. Recognize that there is no guarantee that it would indeed occur once every 10 years. In fact, it could happen 3 years in a row and then not again for 50 years. On the average, however, a ten-year recurrence interval is a reliable estimate for this example.

WATER TRANSMISSION

Water can be transported from either a ground or surface supply source directly to a community or if water quality considerations indicate, initially to a water treatment facility by different types of conduits including:

- Pressure conduits: tunnels, aqueducts, and pipelines
- Gravity-flow conduits: grade tunnels, grade aqueducts, and pipelines

The location of the well field or river reservoir defines the length of the conduits while the topography indicates whether the conduits are designed to carry the water in open-channel flow or under pressure. The profile of a water supply conduit must generally follow the hydraulic grade line to take advantage of the forces of gravity and thus minimize pumping costs.

Service reservoirs are also necessary in the transmission system to help level out peak demands. In practice, intermediate reservoirs close to the city or water towers are sized to meet three design constraints:

- hourly fluctuations in water consumption within the service area
- short-term shutdown of the supply network for servicing
- back-up water requirements to control fires

These distribution reservoirs are most often constructed as open or covered basins, elevated tanks, or, in the past, standpipes. If the service reservoirs are adequately designed to meet these capacity considerations, then the supply conduits leading to them generally must only be designed to carry approximately 50 percent in excess of the average daily demand of the system or subsystem.

CONCLUSION

As the hydrologic cycle indicates, water is a renewable resource because of the driving force of the sun. Thus the earth is not running out of water, it may just be running out of clean water in general and sufficient water in isolated areas as climatic changes take place.

Both groundwater and surface water supplies are available, though to varying degrees, across the face of the earth. The trick is to develop these supplies using sound engineering design procedures for wells and dams, and to protect these supplies using sound engineering and ethical judgments. In the next chapter, we address methods of preparing and treating water for distribution and consumption once the supply has been provided.

PROBLEMS

4.1 A storage reservoir is needed to ensure a constant flow of 20 cfs to a city. The monthly stream flow records are:

Month	J	F	M	A	M	J	J	A	S	O	N	D
ft^3 of water $\times$ 10^6	60	70	85	50	40	25	55	85	20	55	70	90

Calculate the storage requirement.

4.2 If a faucet drips at a rate of 2 drops per second and it takes 25,000 drops to equal 1 gallon of water, how much water is lost each day? Each year? If water costs $1.20/1000 gal, how long would the lost water take to equal the cost of fixing the leak if you did it yourself (free labor!) for 10¢ in parts.

4.3 Icebergs from the South Pole have been suggested as a source of potable water for equatorial Africa. If the technology were available to drag the ice from its source (the South Pole) to its demand area (equatorial Africa), how large an ice cube would be necessary to supply 10,000 people for one year? State all assumptions about travel time and rate of melting.

4.4 Assume you are in a rowboat floating in the middle of a reservoir. Next to you on the seat is a large rock approximately 1 ft^3. You throw the rock into the water; the rock quickly sinks to the bottom. Does the water level in the reservoir, as measured at the shoreline, go up, go down, or remain unchanged?

Chapter 5

Water Treatment

Many aquifers and isolated surface waters are high in water quality and may be pumped from the supply and transmission network directly to any number of end uses, including human consumption, irrigation, industrial processes, or fire control. However, such clean water sources are the exception to the rule in many regions of the nation, particularly regions with dense populations or regions that are heavily agricultural. Here, the water supply must receive varying degrees of treatment prior to distribution.

Impurities enter the water as it moves through the atmosphere, across the earth's surface, and between soil particles in the ground. These background levels of impurities are often supplemented by human activities. Chemicals from industrial discharges and pathogenic organisms of human origin, if allowed to enter the water distribution system, can cause health problems. Excessive silt and other solids can make the water both unsightly and aesthetically unpleasing. Water can be contaminated by many routes. For example, heavy metal pollution, including lead, zinc, and copper, can be caused by corrosion of the very pipes that carry the water from its source to the consumer.

The method and degree of water treatment are important considerations for environmental engineers. Generally speaking, the characteristics of raw water determine the method of treatment. Because most public supply systems are relied on for drinking water, as well as industrial and fire consumption, the highest level of use, human consumption, defines the degree of treatment. Thus we focus only on treatment technologies that produce potable water.

A typical water treatment plant is diagrammed in Figure 5-1. These plants are designed to remove odors, color, and turbidity as well as bacteria and other contaminants from surface water. Raw surface water

77

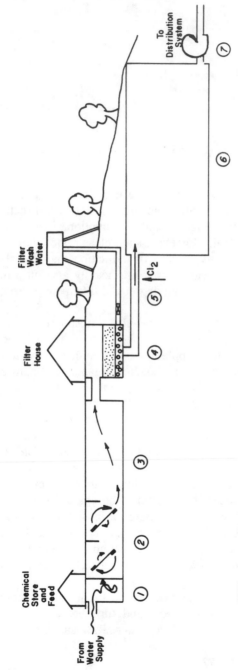

Figure 5-1   Movement of water through a typical water treatment plant.

KEY:

1: Chemical mixing basin ⎤ Coagulation and Flocculation
2: Flocculation basin ⎦
3: Settling tank
4: Rapid sand filter
5: Disinfection with chlorine
6: Clean water storage basin   (Clear well)
7: Pump

entering a water treatment plant usually has significant turbidity caused by tiny colloidal clay and silt particles. These particles have a natural electrostatic charge that keeps them continually in motion and prevents them from colliding and sticking together. Chemicals such as alum (aluminum sulfate) are added to the water, first to neutralize the charge on the particles and then to aid in making the tiny particles ''sticky'' so they can coalesce and form large particles called *flocs*. This process is called *coagulation* and *flocculation* and is represented by Stages 1 and 2 in Figure 5-1.

COAGULATION AND FLOCCULATION

A simple but not altogether satisfactory explanation of flocculation assumes charged particles that move through the water. A negatively charged solid particle will attract ions of the opposite charge—positive— that form a layer over the particle's surface. These charges, in turn, attract negative charges to form an outer layer as shown in Figure 5-2. The net effect is that particles surrounded by these charged layers repel each other: Energy barriers exist between them. The objective of coagulation or flocculation is the reduction of this energy barrier.

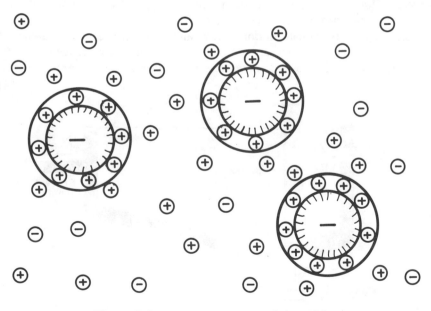

Figure 5-2   Charges on suspended particles.

One means of accomplishing this end is to add trivalent cations to the water. These ions would snuggle up to the negatively charged particle and, because they possess a stronger charge, displace the monovalent cations. The effect of this would be to reduce the net negative charge and thus lower the repulsive force. In this condition, the particles will not repel each other and, upon colliding, will stick together. A stable colloidal suspension has thus been made into an unstable colloidal suspension.

The usual source of trivalent cations in water treatment is alum (aluminum sulfate). Alum has an additional advantage in that some fraction of the aluminum may form aluminum oxides/hydroxides, represented simply as:

$$Al^{+++} + 3OH^- \rightarrow AlOH_3\downarrow$$

These complexes are sticky and heavy and will greatly assist in the clarification of the water in the settling tank if the unstable colloidal particles can be made to come into contact with the floc. This process is enhanced through the operation known as flocculation.

SETTLING

When the flocs have been formed they must be separated from the water. This is invariably done in gravity settling tanks that simply allow the heavier-than-water particles to settle to the bottom. Settling tanks are designed to approximate uniform flow and to minimize all turbulence. Hence the two critical elements of a settling tank are the entrance and exit configurations. Figure 5–3 shows one type of entrance and exit config-

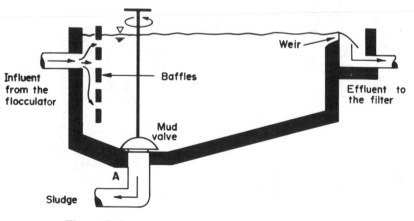

Figure 5–3   Schematic of a common settling tank.

uration used for distributing the flow entering and leaving the water treatment settling tank.

Alum sludge is not highly biodegradable and thus will not decompose at the bottom of the tank. After some time, usually several weeks, the accumulation of alum sludge at the bottom of the tank is such that it has to be removed. Typically, the sludge exits through a *mud valve* at the bottom and is either wasted into a sewer or to a sludge holding/drying pond.

The water leaving a settling tank is essentially clear. The polishing is performed with a rapid sand filter.

FILTRATION

In Chapter 4 we discuss movement of water into the ground and through soil particles, and allude to the cleansing action the particles have on contaminants in the water. Picture the extremely clear water that bubbles up from "underground streams" as spring water. The soil particles definitively help filter the groundwater, and through the years environmental engineers have learned to apply this natural process in water treatment and supply systems and developed what we now know as the *rapid sand filter*. The actual process of separating impurities from a carrying liquid by rapid sand filtration involves two phases: filtration and backwashing.

A slightly simplified version of the rapid sand filter is illustrated in a cut-away in Figure 5-4. Water from the settling basins enters the filter and seeps through the sand and gravel bed, through a false floor, and out into a clear well that stores the finished water. During filtration valves A and C are open.

Eventually the rapid sand filter becomes clogged and must be cleaned. This cleaning is performed hydraulically. The operator first shuts off the flow of water to the filter (closing valves A and C), then opens valves D and B, which allow wash water (clean water stored in an elevated tank or pumped from the clear well) to enter below the filter bed. This rush of water forces the sand and gravel bed to expand and jolts individual sand particles into motion, rubbing against their neighbors. The light colloidal material trapped within the filter is released and escapes with the wash water. After a few minutes, the wash water is shut off and filtration is resumed.

Filter beds are often classified as single medium, dual media, or trimedia. The latter two are often utilized in the treatment of wastewater because they permit solids to penetrate into the bed, have more storage capacity, and thus increase the required time between backwashings. Also, multimedia filters tend to spread head loss buildup over time and further permit longer filter runs.

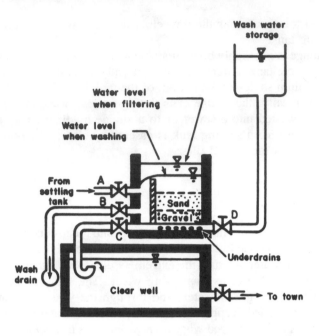

Figure 5-4   Schematic of a rapid sand filter.

DISINFECTION

After filtration, the finished water is often disinfected with chlorine (Step 5 in Figure 5-1). Disinfection is the process of killing the remaining microorganisms in the water some of which may be pathogenic. Chlorine from bottles or drums is fed in correct proportions to the water in order to obtain a desired level of chlorine in the finished water. When chlorine comes in contact with any organic matter including microorganisms, it oxidizes this material and is in turn reduced to inactive chlorides.

Chlorine gas is very soluble in water and rapidly forms hypochlorous acid through hydrolysis:

$$Cl_2 + H_2O \rightleftarrows HOCl + H^+ + Cl^-$$

The hypochlorous acid itself ionizes:

$$HOCl \rightleftarrows OCl^- + H^+$$

At temperatures found in water supply systems, the hydrolysis of chlorine is generally complete in a matter of seconds while the ionization of hypochlorous acid is instantaneous. Both HOCl and OCl⁻ are effective disinfectants and are called *free available chlorine* in water. This free available chlorine kills pathogenic bacteria and thus disinfects the water.

Many water plant operators prefer to maintain a residual of chlorine in the water, thus assuring that, if within the distribution system organic matter such as bacteria, enters the water, there is sufficient chlorine present to eliminate this potential health hazard.

It must be noted that the secondary effects of chlorine addition are poorly understood. Chlorine may combine with synthetic organics in the water to possibly produce dangerous halogenated compounds, and the long-term health effects of such compounds as trihalomethane and chloroform are unknown.

In addition to chlorine, a growing number of municipalities now have fluoride added to the water. At proper concentration fluoride will do much to prevent dental decay in children and young adults.

From the clear well (Step 6 in Figure 5–1), the water is pumped to the distribution system. This is a closed network of pipes, all under pressure. In most cases water is pumped to an elevated storage tank that not only serves to equalize pressures but also provides storage for fires and other emergencies.

CONCLUSION

Water treatment is often necessary if surface water supplies, and sometimes groundwater supplies, are to be available for human consumption. Because the vast majority of cities across the nation use one water distribution system for households, industries, and fire control, large quantities of water often must be available to satisfy the "highest use," which is usually drinking water. Does it make sense to go to the trouble and expense of producing high-quality water only to use it for lawn sprinkling? This is not an easy question, and the problems of future water supply have prompted the serious consideration of dual water supplies: one of high quality and one of a lower quality, perhaps reclaimed from wastewater. Many engineers are convinced that the next major environmental engineering concern will be the availability and production of potable water. The job, therefore, is far from done.

PROBLEMS

5.1 Discuss utilization and disposal options for water treatment sludges collected in the settling tanks following flocculation basins.

5.2 In Figure 5–3, why are there baffles at the entrance? What might the flow look like without baffles?

5.3 In Figure 5–4, it has been found that the first rush of water through a newly cleaned filter carries some solids and should be wasted. Design a piping system to allow for such wastage.

5.4 Suppose that, of the unit operations shown in Figure 5–1, one (and only one) could be duplicated so that a spare unit would always be in reserve. Which unit operator would you select? Why?

Chapter 6

Collection of Wastewater

"The Shambles" is both a street and an area in London and during the eighteenth and nineteenth centuries was a highly commercialized area with meat packing as a major industry. The butchers in those days would throw all of their wastes into the street where it was washed by rainwater into drainage ditches. The condition of the area was so bad that it contributed its name to the English language.

In old cities, drainage ditches like the ones at the Shambles were constructed for the sole purpose of moving stormwater out of the cities. In fact, it was illegal in London to discard human excrement into these ditches. Eventually, these ditches were covered over and became what we now know as *storm sewers*.

As water supplies developed and the use of the indoor water closet increased, the need for transporting domestic wastewaters, called sanitary wastes, became obvious. This was accomplished in one of two ways: (1) discharge of the sanitary wastes into the storm sewers, which then carried both sanitary wastes and stormwater and were known as *combined sewers,* or (2) construction of a new system of underground pipes for removing the wastewater, which became known as *sanitary sewers.*

Newer cities and more recently built (post-1900) parts of older cities almost all have separate sewers for sanitary wastes and stormwater. In this chapter, storm sewer design is not covered in detail since this is discussed in Chapter 9. Emphasis here is on estimating the quantities of domestic and industrial wastewaters and in the design of the sewerage systems to handle these flows.

ESTIMATING WASTEWATER QUANTITIES

The term *sewage* is used here to mean only domestic wastewater. In addition to sewage, however, sewers also must carry

- industrial wastes
- infiltration
- inflow

The quantity of industrial wastes can usually be established by water use records. Alternatively, the flows can be measured in manholes that serve only a specific industry, using a small flow meter in a manhole. Industrial flows often vary considerably throughout the day, and continuous recording is mandatory.

Infiltration is the flow of groundwater into sanitary sewers. Sewers are often placed under the groundwater table, and any cracks in the pipes will allow water to seep in. Infiltration is the least for new, well-constructed sewers and can go as high as 500 m³/km-day (200,000 gal/mi-day). Commonly, for older systems, 700 m³/km-day (300,000 gal/mi-day) is used in estimating infiltration. This flow is of course detrimental since the extra volume of water must go through the sewers and the wastewater treatment plant. It thus makes sense to reduce this as much as possible by maintaining and repairing sewers and keeping sewerage easements clear of large trees that could send roots into the sewers and cause severe damage.

The third source of flow in sanitary sewers is called inflow and represents stormwater that is collected unintentionally by the sanitary sewers. A common source of inflow is a perforated manhole cover placed in a depression so that stormwater flows into the manhole. Sewers laid next to creeks and drainageways, which rise up higher than the manhole elevation or where the manhole is broken, are also a major source. Lastly, illegal connections to sanitary sewers, such as roofdrains, can substantially increase the wet weather flow over the dry weather flow. Commonly, the ratio of dry weather to wet weather flow is between 1:1.2 and 1:4.

Domestic wastewater flows vary with season, day of the week, and hour of the day. Figure 6–1 shows a typical daily flow for a residential area.

The three flows of concern when designing sewers are the average flow, the peak or maximum flow, and the extreme minimum flow. The ratios of average to both the maximum and minimum flows is a function of the total flow since a higher average daily discharge implies a larger community in which the extremes are evened out.

SYSTEM LAYOUT

Sewers that collect wastewater from residences and industrial establishments almost always operate as open channels or gravity flow conduits. Pressure sewers are used in a few places, but these are expensive to main-

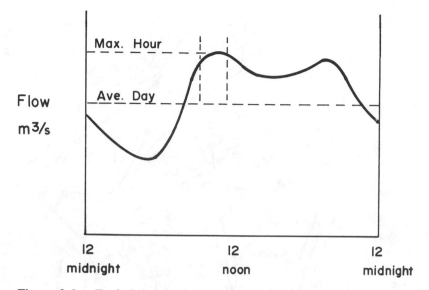

Figure 6-1   Typical dry-weather wastewater flow for a residential area.

tain and are useful only when either there are severe restrictions on water use or the terrain is such that gravity flow conduits cannot be efficiently constructed.

A typical system for a residential area is shown in Figure 6-2. Building connections are usually made with clay or plastic pipe, 6 in. in diameter, to the *collecting sewers,* which commonly run under the street. Collecting sewers are sized to carry the maximum anticipated peak flows without surcharging (filling up) and are commonly made of clay, asbestos, cement, concrete, or cast iron pipe. They discharge in turn into intercepting sewers, known colloquially as *interceptors,* which collect large areas and discharge finally into the wastewater treatment plant.

Collecting and intercepting sewers must be placed at a sufficient grade to allow for adequate velocity during low flows but not so great a grade as to promote excessively high velocities when the flows are at their maximum. In addition, sewers must have manholes, commonly every 120-180 m (400-600 ft) to facilitate cleaning and repair. Manholes are also necessary whenever the sewer changes grade (slope), size, or direction. Typical manholes are shown in Figure 6-3.

In some cases it becomes either impossible or uneconomical to use gravity flow, and the wastewater is pumped. A typical packaged pumping station is shown in Figure 6-4, as its use is indicated on the typical system layout in Figure 6-2.

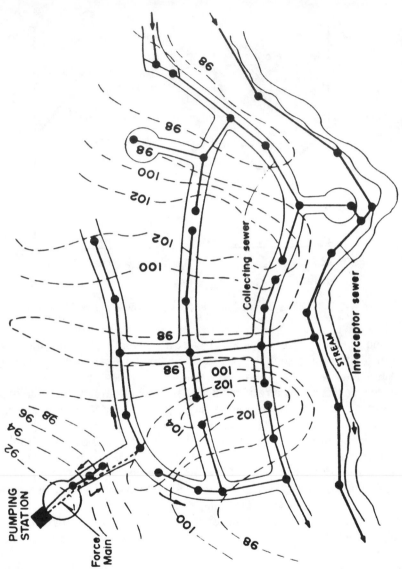

Figure 6–2   Typical wastewater collection system layout [after J. Clark, W. Viessman, and M. Hammer *Water and Sewerage* (New York: IEP, 1977)].

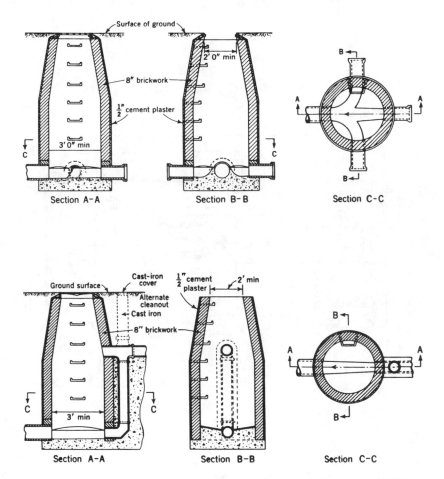

Figure 6–3   Typical manholes used for collecting sewers (courtesy ASCE).

CONCLUSION

"Down the drain!" is both a euphemism and a solution to many nasty cleanup problems. The sewerage systems must be designed to be able to handle almost anything people can pour (or cram) down the drain. Sewerage systems, however, are not ultimate disposal facilities. Sewers must empty somewhere. The treatment of the material going "down the drain" is the topic of the next chapter.

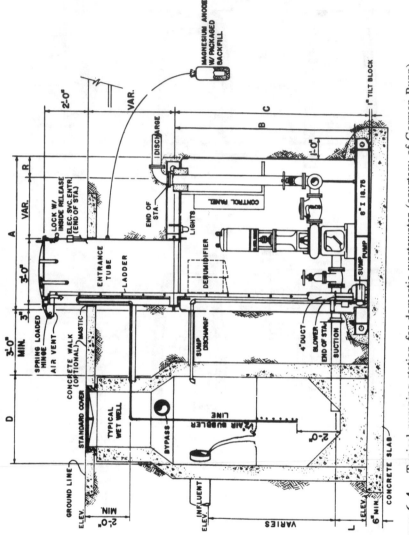

Figure 6-4   Typical pumping station for domestic wastewater (courtesy of Gorman-Rupp).

PROBLEMS

6.1 A community of 100,000 produces an average dry weather wastewater flow of 120 gal/capita/day. What would be the range of the expected wet weather flow?

6.2 In sewer hydraulics, what is meant by "maximum hour"?

6.3 Using your campus road map, design a sewerage system, placing manholes and pumping stations as needed.

Chapter 7
Wastewater Treatment

As civilization developed and cities grew, domestic sewage and industrial wastes were eventually discharged into drainage ditches and sewers, and the entire contents commonly emptied into the nearest watercourse. In the case of major cities, this discharge was often sufficient to destroy even a large body of water. As Samuel Taylor Coleridge observed:

> In Köln, a town of monks and bones
> And pavements fanged with murderous stones
> And rags, and bags, and hideous wenches;
> I counted two and seventy stenches,
> All well defined, and several stinks!

> Ye Nymphs that reign o'er sewers and sinks,
> The river Rhine, it is well known,
> Doth wash your city of Cologne;[1]
> But tell me Nymphs! What power divine
> Shall hence forth wash the river Rhine?

During the nineteenth century, the river Thames was so grossly polluted that the House of Commons had to have rags soaked in lye stuffed into the cracks in the windows of Parliament to reduce the stench.

Beginning with the pioneering work in the United States and England, sanitary engineering technology eventually developed to the point where it became economically, socially, and politically feasible to treat the wastewater so as to reduce its adverse impact on watercourses. In this chapter, this technology is reviewed, beginning with the simplest (and earliest) treatment systems and concluding with a description of the most advanced systems in use today. The discussion begins, however, by reviewing the characteristics of wastewaters that make their disposal difficult and

---

[1]Cologne and Köln are the same city. The German name is Köln; the French, Cologne.

93

showing why wastewater disposal cannot always be done onsite and why sewers and centralized treatment plants become necessary.

WASTEWATER CHARACTERISTICS

The discharges in a sanitary sewerage system are comprised of domestic wastewater, industrial discharge, and infiltration. Infiltration, of course, is only a problem in that it adds to the total volume of wastewater and seldom is it directly a cause of concern in wastewater disposal. Industrial wastes are of course another matter. These discharges vary widely with the size and type of industry and the amount of treatment applied by the industry prior to discharge into public sewers. In the United States, the trend has been toward a higher degree of pretreatment (see Chapter 10), prompted in part by regulations limiting discharges and by the imposition of local sewers surcharges. The latter are charges levied by a community to help pay for the extra treatment that an unusual discharge would require.

Commonly, the problems with industrial discharges are not BOD and suspended solids, both of which can be readily reduced in a wastewater treatment plant, but chemicals such as toxic metals, radioactive materials, refractory organics, and so forth. Typically, local communities place tight restrictions on such discharges and thus force the industries to pretreat the wastewater prior to discharging to the public sewers.

The third component of municipal wastewaters, domestic sewage, tends to vary substantially over time and from one community to the next. On the average, however, it is instructive to consider "typical" values for the characteristics of domestic wastewater. Some of these are shown in Table 7–1.

Table 7–1.   Characteristics of a Typical Domestic Wastewater

Parameter	Typical Value for Domestic Sewage
Biochemical Oxygen Demand	200 mg/L
Suspended Solids	180 mg/L
Phosphorus	8 mg/L
Organic and Ammonia Nitrogen	40 mg/L
pH	6.8
Chemical Oxygen Demand	500 mg/L
Total Solids	720 mg/L

ONSITE WASTEWATER DISPOSAL

Environmental engineers have been severely (and sometimes rightly) criticized for having a "sewer syndrome"—they want to collect all wastewater and provide the treatment at a large central location. Often this approach does not make much sense.

Consider a situation depicted in Figure 7-1, where two wastewater treatment options are shown—a centralized treatment plant and several smaller plants—all discharging their effluents into the same river. The single large plant obviously must provide extremely good treatment in order to attain acceptable dissolved oxygen levels downstream. On the other hand, the smaller plants could take advantage of the assimilative capacity of the river and would not necessarily have to provide the same high degree of treatment. The logical extension of this idea is to not have *any* treatment plants at all, but dispose of the wastewater onsite with each house or building having its own treatment system.

The original onsite system of course is the pit privy, glorified in song and fable.[2] The privy, still used in camps and other temporary residences, consists simply of a deep (perhaps 2 m or 6 ft) pit into which human excrement is deposited. When a pit fills up, it is covered, and a new pit is dug.

A logical extension of the pit privy idea is a composting toilet that accepts not only human wastes but food waste as well and produces a useful compost. With such a system, wastewaters from other sources such as washing and bathing are discharged separately. Composting toilets are on the market at present. The commercially available ones are odor free and compost all organic waste. There is, however, still considerable resistance to installing an indoor toilet that doesn't flush!

By far the most common onsite sewage disposal method is a septic tank with a tile drain field. As shown in Figure 7-2 a septic tank consists of

---

[2]Probably the most famous literary work on the theme of the outhouse was by James Whitcomb Riley, who penned "The Passing of the Backhouse." A few lines from this epic:

"But when the crust was on the snow
   and the sullen skies were grey,
In sooth the building was no place where
   one could wish to stay.
We did our duties promptly,
   there one purpose swayed the mind.
We tarried not nor lingered long
   on what we left behind,
The torture of that icy seat
   would make a Spartan sob. . ."

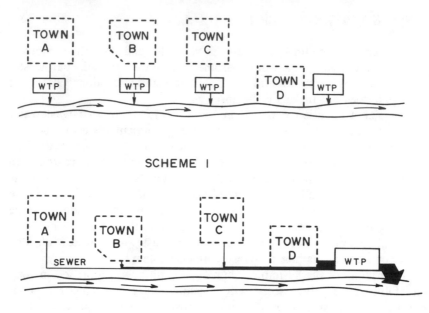

Figure 7-1   Two wastewater treatment options for several small communities.

a concrete box that removes the solids and promotes partial decomposition. The solid particles settle out and eventually fill the tank thus necessitating periodic cleaning. The water overflows into a tile drain field that promotes the seepage of water discharged.

A tile field consists of pipe laid in about a 1-m (3-ft) deep trench, end on end but with short gaps between the pipes. The effluent from the septic tank flows into the tile field pipes and seeps into the ground through these gaps. Alternatively, seepage pits consisting of gravel and sand can be used for promoting the adsorption of the effluent into the ground.

The most important consideration in designing a septic tank and tile field system is the ability of the ground to absorb the effluent. Percolation tests must be made to measure the suitability of the ground for a drain field.

The U.S. Public Health Service and all county and local departments of health have established guidelines for sizing the tile fields or seepage pits. Typical standards are shown in Table 7-2.

Many areas in the United States have soils that percolate poorly, and septic tank/tile field systems are inappropriate. Several options are now

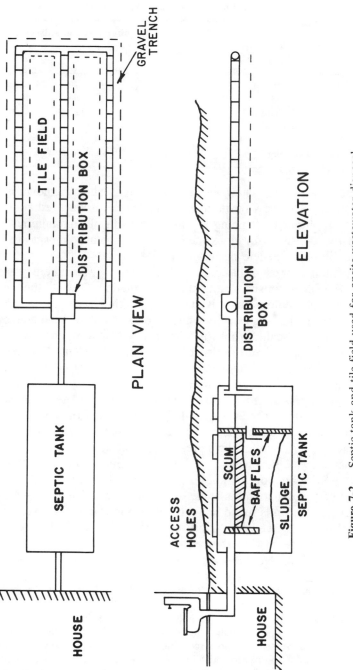

Figure 7-2 Septic tank and tile field used for onsite wastewater disposal.

Table 7–2.   Adsorption Area Requirements for Private Residences

Percolation Rate (in/min)	Required Area of Adsorption Field, per Bedroom (ft^2)
Greater than 1	70
Between 1 and 0.5	85
Between 0.5 and 0.2	125
Between 0.2 and 0.07	190
Between 0.07 and 0.03	250
Less than 0.03	unsuitable ground

available for onsite wastewater disposal, one of which is shown in Figure 7–3. Since 1970, new onsite disposal has been discouraged and in some regions prohibited. Planning for future growth of communities and the legislative demands for assessment of environmental impact are making onsite disposal obsolete in many areas.

In urbanized areas, it has been several centuries since there was enough land available for onsite treatment and percolation. Up until the nineteenth century, this problem was solved by constructing large cesspools or holding basins for the wastewater, which had to be pumped out as they filled up. The public health problems with this system were enormous, as discussed previously (see the Broad Street pump incident, Chapter 2). An obviously better way to move the human waste out of a congested community was to use water as a carrier.

The *water closet,* as it is known in Europe, thus became a standard trapping of our urban society. Actually, the invention of the flushed water closet is mired in controversy. Some credit John Bramah[3] with being the inventor in 1778, while others present convincing arguments that Sir John Harington[4] was the inventor in 1596.[5] The latter argument is strengthened by Sir John's original description of the device although there is no record of him donating his name to the invention. The first recorded use of that euphemism is found in the regulation at Harvard University where in 1735 it was decreed that "No Freshman shall go to the Fellows' John."

The wide use of waterborne wastewater disposal, however, caused another problem. Now the wastes of a community were all concentrated

[3] R.S. Kirby, et al. *Engineering in History* (New York: McGraw-Hill Book Company, 1956).

[4] W. Reyburn, *Flushed With Pride* (London: McDonald, 1969).

[5] Sir John, a courtier-poet, installed his invention in his country house at Kelson, near Bath. Queen Elizabeth had one fitted soon afterwards at Richmond Palace. The two books that were written about this innovation bear the strange titles, *A New Discourse on a Stale Subject. Called the Metamorphosis of Ajax,* and *An Anatomie of the Metamorphased Ajax.* Ajax is a play on words; "a jakes" is a closet.

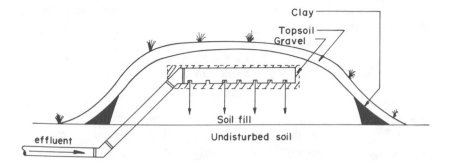

Figure 7-3 Alternative onsite disposal system.

in one place, and a major effort was necessary to clean up the mess. This demand fostered what is known as centralized treatment.

CENTRAL WASTEWATER TREATMENT

The objective of wastewater treatment is to reduce the concentrations of specific pollutants to the level where the discharge of the effluent will not adversely affect the environment. Note that there are two important aspects of this objective. First, wastewater is treated only to reduce the concentrations of selected constituents that would cause harm to the environment or pose a health hazard. Not everything in wastewater is troublesome and thus is not removed. Secondly, the reduction of these constituents is only to some required level. It is obviously technically possible to produce distilled and deionized H_2O from wastewater, but this is not necessary and can in fact be detrimental to the watercourse. Fish and other aquatic organisms cannot survive in distilled water.

For any given wastewater in a specific location, the *degree* and *type* of treatment therefore are variables that require engineering decisions.

Often, the degree of treatment is dictated by the assimilative capacity of the recipient. The procedure by which dissolved oxygen sag curves are drawn is reviewed in Chapter 2. As noted there, the amount of oxygen-demanding materials (BOD) discharged determines how far the dissolved oxygen level will be depressed. If this depression (deficit) is too large, some BOD must be removed in the treatment plant. Thus a certain plant on a given watercourse is required to produce a given quality of effluent. Such an *effluent standard* (discussed more fully in Chapter 10) dictates in large part the type of treatment required.

In order to facilitate the discussion of wastewater treatment, a "typical wastewater" (Table 7-1) will be assumed, and it will be further assumed that the effluent from this wastewater treatment must meet the following effluent standards:

$$BOD_5 \leq 15 \text{ mg}$$
$$SS \leq 15 \text{ mg}$$
$$P \leq 1 \text{ mg}$$

Obviously, other criteria might, in given situations, be important. For example, nitrogen is thought to be the limiting nutrient in estuarine waters, and if the discharge was to be a brackish estuary, the total nitrogen would be an important parameter. In our simplified case, however, we are concerned only with these three constituents.

To further facilitate discussion, the treatment system selected to achieve these effluent levels consists of four major components:

1. *Pretreatment.* Screening and removing large objects such as rat carcasses and chunks of debris.
2. *Primary treatment.* The major objectives are removal of nonhomogenizable solids and homogenization. Primary treatment systems are always physical processes, as opposed to biological or chemical ones.
3. *Secondary treatment.* Designed to remove the demand for oxygen. These processes are commonly biological in nature.
4. *Tertiary treatment.* A name applied to any number of polishing or cleanup processes, one of which is the removal of nutrients such as phosphorus. These processes can be physical (filters), biological (oxidation ponds), or chemical (precipitation of phosphorus).

PRETREATMENT

The most objectionable aspect of discharging raw sewage into watercourses is the floating material. It is only logical, therefore, that *screens* were the first form of wastewater treatment used by communities, and even today, screens are used as the first step in treatment plants. Typical screens, shown in Figure 7–4, consist of a series of steel bars that might be about 2.5 cm (1 in) apart. The purpose of a screen in modern treatment plants is the removal of materials that might damage equipment or hinder further treatment. In some older treatment plants, screens are cleaned by hand, but mechanical cleaning equipment is used in almost all new plants. The cleaning rakes are automatically activated when the screens get sufficiently clogged to raise the water level in front of the bars.

In many plants, the next treatment step is a *comminutor,* a circular grinder designed to grind the solids coming through the screen into pieces about 0.3 cm (0.18 in) or smaller. Many designs are in use; one common design is shown in Figure 7–5.

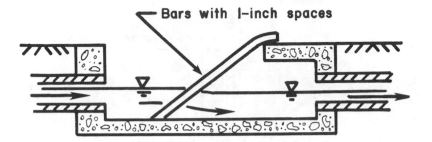

Bars with 1-inch spaces

Figure 7-4 Bar screen used in wastewater treatment. The top picture shows a manually cleaned screen; the bottom picture represents a mechanically cleaned screen (photo courtesy Envirex).

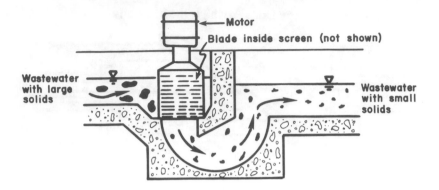

Figure 7-5   A comminutor used to grind up large solids.

The third treatment step involves the removal of grit or sand. This is necessary because grit can wear out and damage such equipment as pumps and flow meters. The most common *grit chamber* is simply a wide place in the channel where the flow is slowed down sufficiently to allow the heavy grit to settle out. Sand is about 2.5 times as dense as most organic solids and thus settles much faster than do less dense solids. The objective of a grit chamber is to remove sand and grit without removing the organic material. The latter must be further treated in the plant, but the sand can be dumped as fill without undue odor or other problems.

The bar screen, comminutor, and grit chamber represent typical pretreatment. At this point, the wastewater is only marginally cleaner with only some of the larger solids and grit removed. Primary treatment is the next step.

PRIMARY TREATMENT

Following the grit chamber most wastewater treatment plants have a *settling tank* (Figure 7-6) to settle out as much of the solid matter as possible. Accordingly, the *retention time*[6] is kept long and turbulence is kept to a minimum. The solids settle to the bottom and are removed through a pipe while the clarified liquid escapes over a V-notch weir, a notched steel plate over which the water flows, promoting equal distribution of liquid discharge all the way around a tank. Settling tanks are also known as *sedimentation tanks* and often as *clarifiers*. The settling tank that follows pre-

---

[6]Retention time is the total time an average slug of water will spend in the tank. This theoretical time is calculated as the time required to fill up a tank. For example, if the volume is 100 m³, and the flow rate is 2 m³ min, the retention time is 100/2 = 50 min. Some authors use *detention time* synonymously with retention time.

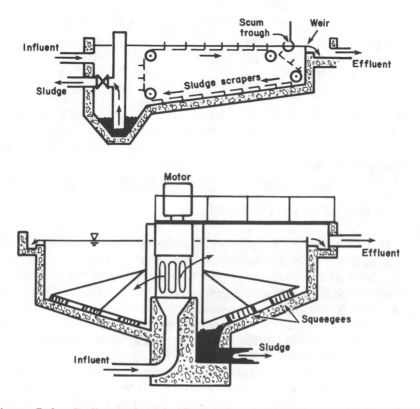

Figure 7-6   Settling tanks (clarifiers). Top drawing shows a rectangular settling tank, and the lower drawing is a circular tank.

liminary treatment such as screening and grit removal is known as a *primary clarifier.* The solids that drop to the bottom of a primary clarifier are removed as *raw sludge,* a name that does not do justice to the undesirable nature of this stuff.

Raw sludge is generally odoriferous and full of water, two characteristics that make its disposal difficult. It must be both stabilized to retard further decomposition and dewatered for ease of disposal. In addition to the solids from the primary clarifier, solids from other processes must similarly be treated and disposed of. The treatment and disposal of wastewater solids (sludge) is an important part of wastewater treatment and is discussed further in Chapter 8.

Primary treatment then is mainly a removal of solids although some BOD is removed as a consequence of the removal of decomposable solids. Typically, the wastewater that was described earlier might now have these characteristics:

	Raw Wastewater	Following Primary Treatment
BOD_5 mg/L	200	150
SS mg/L	180	50
P mg/L	8	7

A substantial fraction of the solids has been removed as well as some BOD and a little P (as a consequence of the removal of raw sludge).

In a typical wastewater treatment plant, this would now move on to secondary treatment.

SECONDARY TREATMENT

The water leaving the primary clarifier has lost much of the solid organic matter but still contains a high demand for oxygen; i.e., it is composed of high-energy molecules that will decompose by microbial action, thus creating a biochemical oxygen demand (BOD). This demand for oxygen must be reduced (energy wasted) if the discharge is not to create unacceptable conditions in the watercourse. The objective of secondary treatment is thus to remove BOD while, by contrast, the objective of primary treatment is to remove solids.

Almost all secondary methods use microbial action to reduce the energy level (BOD) of the waste. The simplest method for putting the microorganisms in the wastewater to work is an aeration device. Farms and feedlots drain wastewater into a pond in which are installed one or more pumps that spray the wastewater into the air, aerating it thoroughly and enhancing the oxidative action of the microorganisms. Industrial effluent whose main objectionable constituent is BOD can be similarly treated in large aeration lagoons.

The concentration of BOD in sanitary sewage is usually so high that aeration alone is insufficient. The first successful means of secondary treatment for sanitary sewage was the trickling filter.

The trickling filter, shown in Figure 7–7, consists of a filter bed of fist-sized rocks over which the waste is trickled. A very active biological growth forms on the rocks, and the organisms obtain their food from the waste stream dripping through the bed of rocks. Air is either forced through the rocks or, more commonly, air circulation is obtained automatically by a temperature difference between the air in the bed and the ambient temperature. In the older filters the waste is sprayed onto the rocks from fixed nozzles. The newer designs use a rotating arm that moves under its own power like a lawn sprinkler, distributing the waste evenly over the entire bed. Often the flow is recirculated thus obtaining

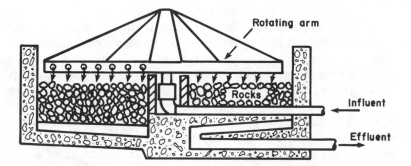

Figure 7-7   Schematic of a trickling filter.

a higher degree of treatment. The name trickling filter is obviously a mis-nomer since no filtration takes place.

Around the turn of the century, when trickling filtration was already firmly established, some researchers began musing about the wasted space in a filter taken up by the rocks. Could the microorganisms not be allowed to float free, and could they not be fed oxygen by bubbling in air? Although this concept was quite attractive, it was not until 1914 that the first workable pilot plant was constructed. It took some time before this process became established as what we now call the *activated sludge system.*

The key to the activated sludge system is the reuse of microorganisms. The system, shown as a block diagram in Figure 7-8, consists of a tank full of waste liquid (from the primary clarifier) and a mass of microorgan-isms. Air is bubbled into this tank (called the *aeration tank*) to provide the necessary oxygen for the survival of the aerobic organisms. The

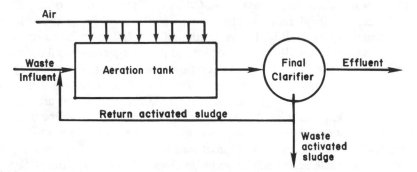

Figure 7-8   Block diagram of the activated sludge system.

microorganisms come in contact with the dissolved organics and rapidly adsorb these organics on their surface. In time, the microorganisms decompose this material to CO_2, H_2O, some stable compounds, and more microorganisms. The production of new organisms is relatively slow, and most of the aeration tank volume is in fact used for this purpose.

Once most of the food has been utilized, the microorganisms are separated from the liquid in a settling tank, sometimes called a *secondary* or *final clarifier.* The liquid escapes over a weir, and can be discharged into the recipient. The separation of microorganisms is an important part of the system. In the settling tanks, the microorganisms exist without additional food and become hungry. They are thus activated; hence the term *activated sludge.*

The settled microorganisms, now known as *return-activated sludge,* are pumped to the head of the aeration tank where they find more food (organics in the effluent from the primary clarifier) and the process starts all over again. The activated sludge process is a continuous operation, with continuous sludge pumping the clean water discharge.

As mentioned earlier, one of the end products of this process is more microorganisms. If none of the microorganisms are removed, their concentration will soon increase to the point where the system is clogged with solids. It is therefore necessary to waste some of the microorganisms, and this *waste activated sludge* must be processed and disposed of. Its disposal is one of the most difficult aspects of waste treatment.

Activated sludge systems are designed on the basis of loading, or the amount of organic matter (food) added relative to the microorganisms available. This ratio is known as the food-to-microorganisms ratio (F/M) and is a major design parameter. Unfortunately it is difficult to measure either F or M accurately, and engineers have approximated these by BOD and the suspended solids in the aeration tank respectively. The combination of the liquid and microorganisms undergoing aeration is known (for some unknown reason) as *mixed liquor,* and thus the suspended solids are called *mixed liquor suspended solids* (MLSS). The ratio of incoming BOD to MLSS, the F/M ratio, is also known as the *loading* on the system, calculated as pounds of BOD/day per pound of MLSS in the aeration tank.

If this ratio is low (little food for lots of microorganisms) and the aeration period (retention time in the aeration tank) is long, the microorganisms make maximum use of available food, resulting in a high degree of treatment. Such systems are known as *extended aeration* and are widely used for isolated sources (for example, motels, small developments). An added advantage of extended aeration is that the ecology within the aeration tank is quite diverse and little excess biomass is created, resulting in little or no waste-activated sludge to be disposed of—a significant saving in operating costs and headaches.

Table 7-3. Loadings and Efficiencies of Activated Sludge Systems

	Loading		Aeration Period (hr)	Efficiency of BOD Removal (%)
	$\dfrac{F}{M} =$	$\dfrac{\text{lb BOD/day}}{\text{lb MLSS}}$		
Extended Aeration		0.05–0.2	30	95
Conventional		0.2–0.5	6	90
High Rate		1–2	4	85

Example 7.1

The BOD_5 of the liquid from the primary clarifier is 120 mg/L at a flow rate of 0.05 mgd. The aeration tank is 20 × 10 × 20 ft, and the MLSS = 2000 mg/L. Calculate the F/M ratio.

$$\text{lb BOD} = 120 \text{ mg/L} \times 0.05 \text{ mgd} \times 8.34 = 50 \text{ lb/day}$$

(see Chapter 3 for discussion of how to convert from flow and concentration to pounds)

$$\text{lb MLSS} = (20 \times 10 \times 20)\text{ft}^3 \times 2000 \text{ mg/L} \times 3.83 \text{ L/gal}$$
$$\times 7.481 \text{ gal/ft}^3 \times 2.20 \times 10^{-6} \text{ lb/mg} = 504 \text{ lb}$$

$$\text{F/M} = \frac{50}{504} = 0.1 \; \frac{\text{lb BOD/day}}{\text{lb MLSS}}$$

At the other extreme is the "high-rate" system where the aeration periods are very short (thus saving money by building smaller tanks) and the treatment efficiency is lower. The efficiencies and F/M ratios for the three types of activated sludge systems are shown in Table 7-3.

When the microorganisms first come in contact with the food, the process requires a great deal of oxygen. Accordingly, the dissolved oxygen level in the aeration tank drops immediately after the point at which the waste is introduced. If DO levels are measured over the length of a tank, extremely low concentrations are often found at the influent end of the aeration tank. These low levels of DO can be detrimental to the microbial population. Accordingly, two variations of the activated sludge treatment have found some use: *tapered aeration* and *step aeration* (Figure 7-9). The former method consists of blasting additional air where needed, while step aeration involves the introduction of the waste at several locations thus evening out the initial oxygen demand.

The third modification is *contact stabilization,* or *biosorption,* a process

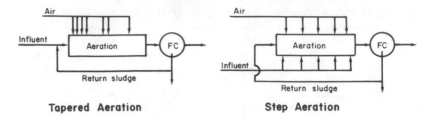

Figure 7-9   Tapered and step aeration schematics.

in which the sorption and bacterial growth phases are separated by a settling tank. The advantage is that the growth can be achieved at high solids concentrations, thus saving tank space. Many existing activated sludge plants can be converted to biosorption plants when tank volume limits treatment efficiency. Figure 7-10 is a diagram of the biosorption process.

The two principal means of introducing sufficient oxygen into the aeration tank are by bubbling compressed air through porous diffusers or beating air in mechanically. Both diffused air and mechanical aeration are shown in Figure 7-11.

The success of the activated sludge system depends on many factors. Of critical importance is the separation of the microorganisms in the final clarifier. The microorganisms in the system are sometimes very difficult to settle out, and the sludge is said to be a *bulking sludge*. Often this condition is characterized by a biomass comprised almost totally of filamentous organisms that form a kind of lattice structure with the filaments and refuse to settle.[7]

Treatment plant operators should keep a close watch on settling characteristics because a trend toward poor settling can be the forerunner of a badly upset (and hence ineffective) plant. The settleability of activated sludge is most often described by the sludge volume index (SVI), which is determined by measuring the mL of volume occupied by a sludge after settling for 30 minutes in a 1-liter cylinder, and calculated as

$$SVI = \frac{(\text{volume of sludge after 30 min, in mL}) \times 1000}{\text{mg/L of suspended solids}}$$

Example 7.2

A sample of mixed liquor was found to have SS = 4000 mg/L and, after settling for 30 minutes in a 1-liter cylinder, occupied 400 mL. Calculate the SVI.

$$SVI = \frac{400 \times 1000}{4000} = 100$$

[7]You can picture this as filling a glass with cotton balls then pouring water in it. The cotton is simply not heavy enough to settle to the bottom of the glass.

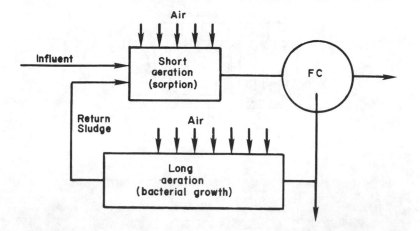

Figure 7–10   The biosorption modification of the activated sludge process.

SVI values below 100 are usually considered acceptable with SVI greater than 200 defined as badly bulking sludges. The causes for poor settling (high SVI) are not always known, and hence the solutions are elusive. Wrong or variable F/M ratios, fluctuations in temperature, high concentrations of heavy metals, and deficiencies in nutrients in the incoming wastewater have all been blamed for bulking. Cures include chlorination, changes in air supply, and dosing with hydrogen peroxide (H_2O_2) to kill the filamentous microorganisms.

When the sludge does not settle, the return-activated sludge becomes thin (low suspended solids concentration), and thus the concentration of microorganisms in the aeration tank drops. This results in a higher F/M ratio (same food input but fewer microorganisms) and a reduced BOD removal efficiency.[8]

Secondary treatment of wastewater then usually consists of a biological step such as activated sludge, which removes a substantial part of the BOD and the remaining solids. Looking once again at the typical wastewater, we now have the following approximate water quality:

	Raw Wastewater	*Following Primary Treatment*	*Following Secondary Treatment*
BOD$_5$ mg/L	200	150	15
SS mg/L	180	50	15
P mg/L	8	7	6

The effluent in fact meets our previously established effluent standards for BOD and SS. Only the phosphorus remains high. The removal of inor-

---

[8]You can think of the microorganisms as workers in an industrial plant. If the total number of workers is decreased, the production is cut. Similarly if fewer microorganisms are available, less work is done.

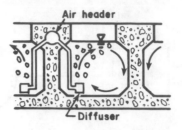

Diffused Aeration

Figure 7–11   Activated systems with diffused aeration and
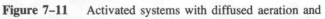

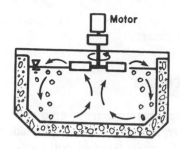

Mechanical Aeration

mechanical (surface) aeration (photos courtesy Envirex).

organic chemicals like phosphorus is accomplished in tertiary (or advanced) wastewater treatment.

TERTIARY TREATMENT

Primary and biological treatments make up the conventional wastewater treatment plant. However, secondary treatment plant effluents still contain a significant amount of various types of pollutants. Suspended solids, in addition to contributing to BOD, can settle out in streams and form unsightly mud banks. The BOD, if discharged into a stream with low flow, can still cause damage to aquatic life by depressing the DO. Neither primary nor secondary treatment is effective in removing phosphorus and other nutrients or toxic substances. Suspended solids can be effectively removed by a sandfilter such as those used in water treatment plants.

For BOD removal after secondary treatment, by far the most popular advanced treatment method is the polishing pond, often called the *oxidation pond*. This is essentially a hole in the ground, a large pond used to confine the plant effluent before it is discharged. Such ponds are designed to be aerobic; hence light penetration for algal growth is important, and a large surface area is needed. The reactions occurring within an oxidation pond are depicted in Figure 7-12. Oxidation ponds are sometimes used as the only treatment step if the waste flow is small and the pond area is large.

Activated carbon adsorption is another method of BOD removal, and this process has the added advantage that inorganics as well as organics are removed. The mechanism of adsorption on activated carbon is both chemical and physical with tiny crevices catching and holding colloidal and smaller particles. An activated carbon column is a completely enclosed tube with dirty water pumped up from the bottom and clear water exiting at the top. As the carbon becomes saturated with various materials, it must be removed from the column and regenerated or cleaned. Removal is often continuous with clean carbon being added at the top of the column. The cleaning or regeneration is usually done by heating the carbon in the absence of oxygen. A slight loss in efficiency is noted with regeneration, and some virgin carbon must always be added to ensure effective performance.

Nitrogen removal can be accomplished by first treating the waste thoroughly enough to produce nitrate ions. This usually involves longer detention times in secondary treatment during which bacteria such as *Nitrobacter* and *Nitrosomonas* convert ammonia nitrogen to NO_3^-, a process called *nitrification*. These reactions are

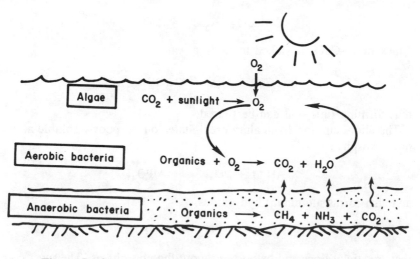

Figure 7–12 Simplified reactions within an oxidation pond.

$$2NH_4^+ + 3O_2 \xrightarrow{\text{\textit{Nitrosomonas}}} 2NO_2^- + 2H_2O + 4H^+$$

$$2NO_2^- + O_2 \xrightarrow{\text{\textit{Nitrobacter}}} 2NO_3^-$$

These reactions are slow and thus require long retention times in the aeration tank, as well as sufficient dissolved oxygen. The kinetics constants for these reactions are low with very low yields so that the net sludge production is limited, making washout a constant danger.

Once the ammonia has been converted to nitrate, it can be reduced by a broad range of facultative and anaerobic bacteria such as *Pseudomonas*. This reduction, called *denitrification*, requires a source of carbon, and methanol (CH_3OH) is often used for that purpose.

$$6NO_3^- + 2CH_3OH \rightarrow 6NO_2^- + 2CO_2\uparrow + 4H_2O$$

$$6NO_2^- + 3CH_3OH \rightarrow 3N_2\uparrow + 3CO_2\rightarrow + 3H_2O + 6OH^-$$

Phosphate removal is usually accomplished chemically. The most popular chemicals used for phosphorus removal are lime $Ca(OH)_2$, and alum, $Al_2(SO_4)_3$. The calcium ion, in the presence of high pH, will combine with phosphate to form a white, insoluble precipitate called calcium hydroxyapatite, which is settled out and removed. Insoluble calcium carbonate is also formed and removed and can be recycled by burning in a furnace.

$$CaCO_3 \xrightarrow{\Delta} CO_2 + CaO$$

Quick lime, CaO, is slaked by adding water,

$$CaO + H_2O \rightarrow Ca(OH)_2$$

thus forming lime that can be reused.

The aluminum ion from alum precipitates out as poorly soluble aluminum phosphate,

$$Al^{+++} + PO_4^{---} \rightarrow AlPO_4\downarrow$$

and also forms aluminum hydroxides,

$$Al^{+++} + 3OH^- \rightarrow Al(OH)_3\downarrow$$

that are sticky flocs and help to settle out the phosphates. The most common point of alum dosing is in the final clarifier.

The amount of alum required to achieve a given level of phosphorus removal depends on the amount of phosphorus in the water as well as on other constituents. The sludge produced can be calculated using stoichiometric relationships.

Using such a technique as alum precipitation of phosphorus (and a bonus of more efficient SS and BOD removal due to the formation of $Al(OH)_3$), we now have attained our effluent goal:

	Raw Wastewater	Following Primary Treatment	Following Secondary Treatment	Following Tertiary Treatment
BOD_5 mg/L	200	150	15	10
SS mg/L	180	50	15	10
P mg/L	8	7	6	0.5

An alternative to high-technology advanced wastewater treatment systems is spraying secondary effluent on land and allowing the soil microorganisms to degrade the remaining organics. Such systems, known as *land treatment,* have been employed for many years in Europe but only recently have been used in North America. They appear to represent a reasonable alternative to complex and expensive systems especially for smaller communities.

Probably the most promising land treatment method is irrigation. Commonly, from 1000 to 2000 hectares of land are required for every 1 m³/sec of wastewater flow, depending on the crop and soil. Nutrients

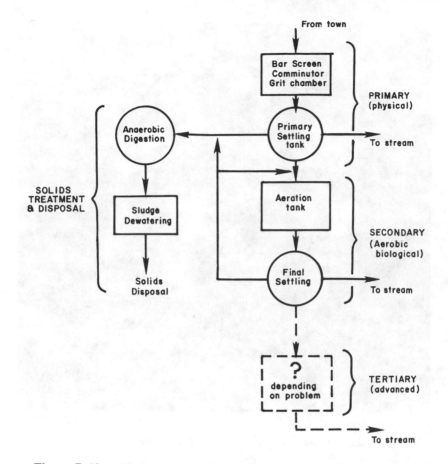

Figure 7-13 Block diagram of a complete wastewater treatment plant.

such as N and P remaining in the secondary effluent are of course beneficial to the crops.

CONCLUSION

A typical wastewater treatment plant is shown schematically in Figure 7-13. We have discussed primary, secondary, and tertiary treatment. The treatment and disposal of the solids removed from the liquid stream deserves special attention, and the topic is covered in the next chapter.

An aerial view of a typical wastewater treatment plant is shown in Figure 7-14. If such plants are well operated, the effluents are often much less polluted than the stream into which they are discharged.

Figure 7–14   Typical wastewater treatment plant.

But not all plants perform that well. The sad fact is that many of the existing wastewater treatment plants are only marginally effective in controlling water pollution, and much of the blame can be placed on plant operation.

The operation of a modern wastewater treatment plant is a complex and demanding job. Unfortunately, operators have historically been considered as being at the bottom of the totem pole both in terms of pay and community stature, and in years past few qualified people were willing to make plant operation a career. Municipalities were forced to use whatever help was available, which often resulted in poor operation.

States now require licensing of treatment plant operators, and their pay and social stature has greatly improved. This is a welcome change for it makes little sense to entrust unqualified and unreliable workers with the operation of multimillion-dollar facilities. Present-day operators are typically highly trained and motivated to make their plants perform to their designed capabilities.

Wastewater treatment thus requires first the proper design of the plant and second the proper plant operation. One without the other is a waste of money.

PROBLEMS

7.1 The following data were reported on the operation of a wastewater treatment plant:

	Influent (mg/L)	Effluent (mg/L)
BOD_5	200	20
SS	220	15
P	10	0.5

a. What percent removal was experienced for each of these?

b. What kind of treatment plant would produce such an effluent? Draw a block diagram showing the treatment steps.

7.2 Describe the condition of a primary clarifier one day after the raw sludge pumps broke down.

7.3 One operational problem with trickling filters is "ponding," the excessive growth of slime on the rocks and subsequent clogging of the spaces so that the water no longer flows through the filter. Suggest some cures for the ponding problem.

7.4 One problem with sanitary sewers is illegal connections. Suppose a family of four, living in a home with a roof area of 70×40 ft, connects the roof drain to the sewer. For a typical rain of 1 in/hr, what percent increase will there be in the flow from their house over the "dry weather" flow (assumed at 50 gal/capita/day)?

7.5 Suppose an industry decided to build a wastewater treatment plant and hired an engineer to design it for them. One of the first steps would be to sample the waste and run some analyses to determine what its characteristics are. If you had to specify the tests to be run, what would you choose as the five most important wastewater parameters of interest? Name the five and state why you want to know these values.

7.6 The influent and effluent data for a secondary treatment plant are:

	Influent (mg/L)	Effluent (mg/L)
BOD	200	20
Suspended solids	200	100
Total phosphorus	10	8

Calculate the removal efficiencies. What do you think might be wrong with the plant?

7.7 Draw block diagrams of the unit operations necessary to treat the following wastes to effluent levels of BOD_5 = 20 mg/L, SS = 20 mg/L, P = 1 mg/L.

Waste	BOD_5 (mg/L)	SS (mg/L)	P (mg/L)
Domestic	200	200	10
Chemical industry	40,000	0	0
Pickle cannery	0	300	1
Fertilizer mfg.	300	300	200

7.8 If the mL of settled sludge (30 min) was 300 mL for both before and during a bulking problem, and the SVI was 100 and 250, what was the MLSS before and after the bulking problem?

7.9 If you conducted a percolation test and discovered that in 30 min the water level dropped 5 in, what size percolation field would you need for a two-bedroom house? What size would you need if it dropped 0.5 in?

7.10 A 1-mgd conventional activated sludge plant has an influent BOD_5 of 200 mg/L. The primary clarifier removes 30 percent of that BOD. The three aeration tanks are each 20 × 20 × 100 ft. What MLSS are necessary to attain 90 percent BOD removal in the plant?

7.11 An aeration system with a hydraulic retention time of 2.5 hr receives a flow of 0.2 mgd at a BOD of 150 mg/L. The suspended solids in the aeration tank are 4000 mg/L. The effluent BOD is 20 mg/L and effluent suspended solids are 30 mg/L. Calculate the F/M ratio (food to microorganism) for this system.

7.12 The MLSS in an aeration tank are 4000 mg/L. The flow from the primary settling tank is 0.2 m³/sec and the return sludge flow is 0.1 m³/sec. What must the return sludge suspended solids concentration be to maintain the 4000 mg/L MLSS?

7.13 A transoceanic flight on a Boeing 747 with 430 persons on board takes 7 hours. Estimate the weight of the water necessary to flush the toilets if each flush required 2 gallons. Make any assumptions necessary. What fraction of the total payload (people) would the flush water represent? How could you reduce this weight? (The railroad system, for obvious reasons, is illegal.)

7.14 A family of four wants to build a house on a lot for which the percolation test results show 1.0 mm/min. The county requires a septic tank hydraulic retention time of 24 hr. Find the volume of the tank required and the area of the tile field. Sketch the system including all dimensions.

LIST OF SYMBOLS

BOD = biochemical oxygen demand, mg/L
F = food (BOD), mg/L
M = microorganisms (SS), mg/L
MLSS = mixed liquor suspended solids, mg/L
P = phosphorus, mg/L

Chapter 8

Sludge Treatment and Disposal

The field of wastewater treatment engineering is littered with unique and imaginative processes for achieving high degrees of waste stabilization at attractive costs. However, few of these "wonder plants" have proven themselves in practice, and quite often the problem has been the inattention to the sludge problem. Drawing a flow diagram with a little arrow labeled "SLUDGE TO DISPOSAL" has often been the total extent of the consideration for solids handling, treatment, and disposal.

In the past few years the fact that sludge treatment and disposal accounts for over 50 percent of the treatment costs in a typical secondary plant has prompted a renewed interest in this none-too-glamorous, but essential, aspect of wastewater treatment.

This chapter is devoted to the problem of sludge treatment and disposal. The sources and quantities of sludge from various types of wastewater treatment systems are examined first followed by a definition of sludge characteristics. Such solids concentration techniques as thickening and dewatering are discussed next, concluding with consideration for ultimate disposal.

SOURCES OF SLUDGE

The first source of sludge is the suspended solids that enter the treatment plant and are partially removed in the primary settling tank or clarifier. Commonly about 60 percent of the suspended solids become *raw primary sludge,* which is highly putrescible and very wet (about 96 percent water).

The removal of BOD is basically a method of wasting energy, and secondary wastewater treatment plants are designed to reduce this high-energy material to low-energy chemicals. This process is typically accomplished by biological means, using microorganisms (the "decomposers" in ecological terms) that use the energy for their own life and procreation. Secondary treatment processes such as the popular activated sludge system are *almost* perfect systems. Their major fault is that the microorganisms convert too little of the high-energy organics to CO_2 and H_2O and too much of it to new organisms. Thus the system operates with an excess of these microorganisms, or *waste-activated sludge*. The mass of waste-activated sludge produced per mass of BOD removed in secondary treatment is known as the *yield*, expressed as kg SS produced/kg BOD removed.

Phosphorus removal processes also invariably end up with excess solids. If lime is used, the calcium carbonates and calcium hydroxyapatites are formed and must be disposed of. Aluminum sulfate similarly produces solids in the form of aluminum hydroxides and aluminum phosphates. Even so-called "totally biological processes" for phosphorous removal end up with solids. The use of an oxidation pond or marsh for phosphorus removal is possible only if some organics (algae, water hyacinths, fish, and so forth) are periodically harvested.

SLUDGE TREATMENT

A great deal of money could be saved and troubles averted if sludge could be disposed of as it is drawn off the main process train. Unfortunately, the sludges have three characteristics that make such a simple solution unlikely: They are aesthetically displeasing and potentially harmful, and they have too much water.

The first two problems are often solved by stabilization, such as *anaerobic* or *aerobic digestion*. The third problem requires the removal of water by either thickening or dewatering. Accordingly, the next three sections cover the topics of stabilization, thickening and dewatering, and ultimate disposal.

Sludge Stabilization

The object of sludge stabilization is to reduce the problems associated with two of the detrimental characteristics listed above, sludge odor and putrescibility, and the presence of pathogenic organisms.

There are two primary means of sludge stabilization:

- aerobic digestion
- anaerobic digestion

Aerobic stabilization is merely a logical extension of the activated sludge system. Waste-activated sludge is placed in dedicated aeration tanks for a very long time, and the concentrated solids are allowed to progress well into the endogenous respiration phase in which food is obtained only by the destruction of other viable organisms. This results in a net reduction in total and volatile solids. Aerobically digested sludges are, however, more difficult to dewater than anaerobic sludges.

The third commonly employed method of sludge stabilization is anaerobic digestion. The biochemistry of anaerobic decomposition of organics is illustrated in Figure 8-1. Note that this is a staged process with the solution of organics by extracellular enzymes being followed by the production of organic acids by a large and hearty group of anaerobic microorganisms known, appropriately enough, as the *acid formers*. The organic acids are in turn degraded further by a group of strict anaerobes called *methane formers*. These microorganisms are the prima donnas of wastewater treatment, getting upset at the least change in their environment.

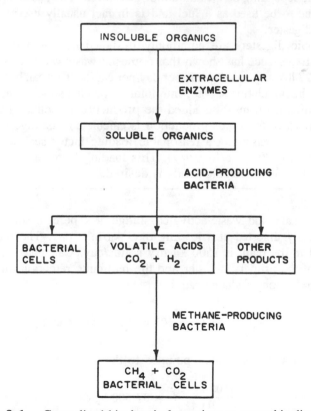

Figure 8-1   Generalized biochemical reactions to anaerobic digestion.

The success of anaerobic treatment thus boils down to the creation of a suitable condition for the methane formers. Since they are strict anaerobes, they are unable to function in the presence of oxygen and very sensitive to environmental conditions such as temperature, pH, and toxins. If a digester goes "sour," the methane formers have been inhibited in some way. The acid formers, however, keep chugging away, making more organic acids. This has the effect of further lowering the pH and making conditions even worse for the methane formers. A sick digester is therefore difficult to cure without massive doses of lime or other antacids.

Many treatment plants have two kinds of digesters—primary and secondary (Figure 8–2). The primary digester is covered, heated, and mixed to increase the reaction rate. The temperature of the sludge is usually about 35 °C (95 °F). Secondary digesters are not mixed or heated and are used for storage of gas and for concentrating the sludge by settling. As the solids settle, the liquid supernatant is pumped back to the main plant for further treatment. The cover of the secondary digester often floats up and down, depending on the amount of gas stored. The gas is high enough in methane to be used as a fuel and is in fact usually used to heat the primary digester.

Anaerobic digesters are commonly designed on the basis of solids loading. Experience has shown that domestic wastewaters contain about 120 g (0.27 lb) suspended solids per day per capita. This can be translated, knowing the population served, into total suspended solids to be handled. To this, of course, must be added the production of solids in secondary treatment. Once the solids production is calculated, the digester volume is estimated by assuming a reasonable loading factor such as 4 kg dry solids/m^3 · day (0.27 lb/ft^3 · day). This loading factor is decreased if a higher reduction of volatile solids is desired.

Example 8.1

Raw primary and waste-activated sludge at 4 percent solids is to be anaerobically digested at a loading of 3 kg/m^3 · day. The total sludge produced in the plant is 1500 kg dry solids/day. Calculate the required volume of the primary digester and the hydraulic retention time.

The production of sludge requires

$$\frac{1500 \text{ kg/day}}{3 \text{ kg/m}^3 \cdot \text{day}} = 500 \text{ mg}^3 \text{ digester volume}$$

The total mass of wet sludge pumped to the digester is

$$\frac{1500 \text{ kg/day}}{0.04} = 37,500 \text{ kg/day}$$

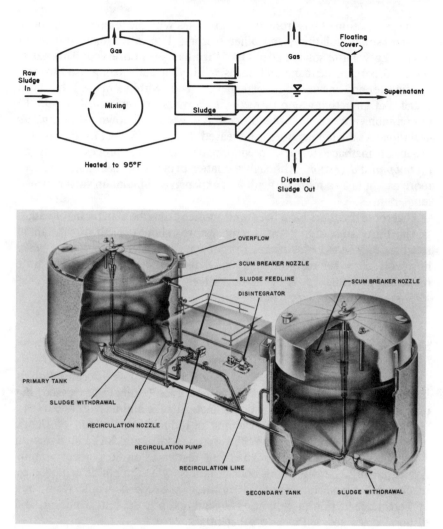

Figure 8–2   Anaerobic sludge digesters (photo courtesy Dorr Oliver Inc.).

and since one liter of sludge weighs about 1 kg, the volume of sludge is 37,500 L/day or 37.5 m³/day and the hydraulic retention time is

$$\bar{t} = \frac{\text{volume}}{\text{flow}} = \frac{500 \text{ m}^3}{37.5 \text{ m}^3/\text{day}} = 13.3 \text{ days}$$

The production of gas from digestion varies with the temperature, solids loading, solids volatility, and other factors. Typically, about 0.6 m³ of gas per kg volatile solids added (10 ft³/lb) has been observed. This gas is about 60 percent methane and burns readily, usually being used to heat the digester and answer additional energy needs within a plant. It has been found that an active group of methane formers operates at 35 °C (95 °F) in common practice, and this process has become known as *mesophilic digestion*. As the temperature is increased, to about 45 °C (115 °F), another group of methane formers predominates; and this process is tagged *thermophilic digestion*. Although the latter process is faster and produces more gas, it is also more difficult and expensive to maintain such elevated temperatures.

Anaerobic digesters have been well studied from the standpoint of pathogen viability since the elevated temperatures should result in substantial sterilization. *Salmonella typhosa* organisms and many other pathogens, however, can survive digestion. Polio viruses similarly survive with little reduction in virulence. An anaerobic digester cannot, therefore, be considered a method of sterilization.

Sludge Thickening

Sludge thickening is a process in which the solids concentration is increased and the total sludge volume is correspondingly decreased, but the sludge still behaves like a liquid instead of a solid.

The advantages of sludge thickening in reducing the volume of sludge to be handled are substantial. With reference to Figure 8–3 a sludge with 1 percent solids thickened to 5 percent results in an 80 percent volume reduction. A 20 percent solids concentration, which might be achieved by mechanical dewatering (discussed in the next section), would result in a 95 percent reduction in volume. The savings in treatment, handling, and disposal costs accrued can be substantial.

Two types of nonmechanical thickening operations are presently in use: the gravity thickener and the flotation thickener. These are not very good names since the latter also uses gravity to separate the solids from the liquid. For the sake of simplicity, however, we will continue to use the two descriptive terms.

A typical gravity thickener looks very much like a circular settling tank (Figure 7–6). The influent, or feed, enters in the middle, and the water moves to the outside, eventually leaving as clear effluent over the weirs. The sludge solids settle as a blanket and are removed out the bottom.

A flotation thickener, shown in Figure 8–4, operates by forcing air under pressure to dissolve in the return flow and releasing the pressure as

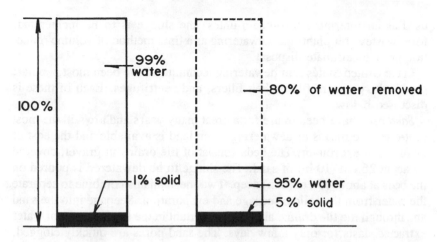

Figure 8-3   Volume reduction due to sludge thickening.

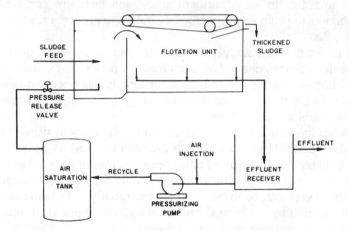

Figure 8-4   Flotation thickener.

the return is mixed with the feed. As the air comes out of solution, the tiny bubbles attach themselves to the solids and carry them upward to be scraped off.

Sludge Dewatering

As defined above, dewatering differs from thickening in that the sludge should behave as a solid after it has been dewatered. Dewatering is seldom used as an intermediate process unless the sludge is to be incinerated. Most wastewater plants use dewatering as a final method of volume reduction prior to ultimate disposal.

In the United States four dewatering techniques have been most popular: sand beds, pressure filters, belt filters, and centrifuges. Each of these is discussed below.

Sand beds have been in use for a great many years and are still the most cost-effective means of dewatering when land is available and the cost of labor is not exorbitant. The beds consist of tile drains in gravel, covered by about 26 cm (10 in) of sand. The sludge to be dewatered is poured on the beds at about 15 cm (6 in) deep. Two mechanisms combine to separate the water from the solids: seepage and evaporation. Seepage into the sand and through the tile drains, although important in the total volume of water extracted, lasts for only a few days. The sand pores are quickly clogged, and all drainage into the sand ceases. The mechanism of evaporation takes over, and this process is actually responsible for the conversion of liquid sludge to solid. In some northern areas sand beds are enclosed under greenhouses to promote evaporation as well as to prevent rain from falling into the beds.

For mixed digested sludge, the usual design is to allow for three months drying time. Some engineers suggest that this period be extended to allow a sand bed to rest for a month after the sludge has been removed. This seems to be an effective means of increasing the drainage efficiency once the sand beds are again flooded.

Raw sludge will not drain well on sand beds and will usually have an obnoxious odor. Hence raw sludges are seldom dried on beds. Raw secondary sludges have a habit of either seeping through the sand or clogging the pores so quickly that no effective drainage takes place. Aerobically digested sludge can be dried on sand, but usually with some difficulty.

If dewatering by use of sand beds is considered impractical, mechanical dewatering techniques must be employed.

The *pressure filter,* shown in Figure 8–5, uses positive pressure to force the water through a filter cloth. Typically, the pressure filters are built as plate-and-frame filters. The sludge solids are captured in the spaces between the plates and frames, which are then pulled apart to allow for sludge cleanout.

The *belt filter,* shown in Figure 8–6, operates both as a pressure filter and by a gravity drainage. As the sludge is introduced onto the moving belt, the free water drips through the belt, but the solids are retained. The belt then moves into the dewatering zone where the sludge is squeezed between two belts. These machines are quite effective in dewatering many different kinds of sludge and are being widely installed in small wastewater treatment plants.

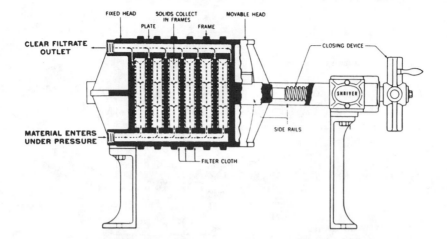

Figure 8–5 Pressure filters (photo courtesy Envirex).

Centrifugation became popular in wastewater treatment only after organic polymers were available for sludge conditioning. The centrifuge most widely used is the solid bowl decanter, which consists of a bullet-shaped body rotating on its long axis. The sludge is placed into the bowl; the solids settle out under about 500 to 1000 gravities (centrifugally applied) and are scraped out of the bowl by a screw conveyor (Figure 8–7).

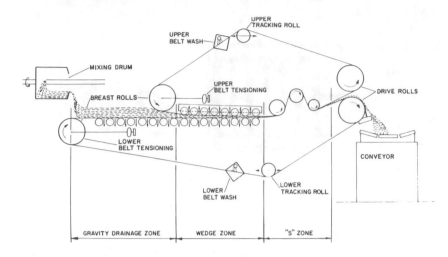

Figure 8–6   Belt filter.

ULTIMATE DISPOSAL

The options for ultimate disposal of sludge are limited to air, water, and land. Strict controls on air pollution complicate incineration and make it expensive, although this certainly is an option. Disposal of sludges in deep

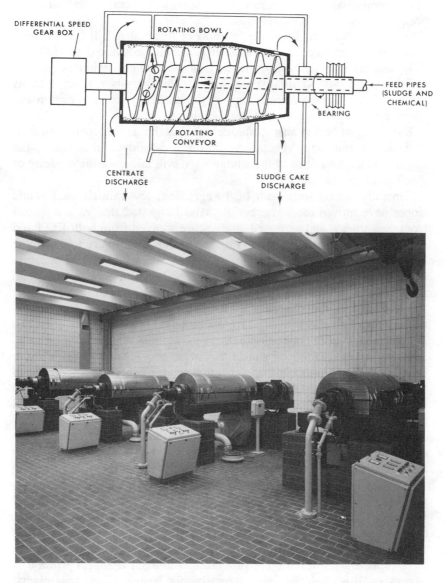

Figure 8-7 Solid bowl centrifuge (photo courtesy Indersoll Rand and I. Krüger).

water (such as oceans) is decreasing due to adverse or unknown detrimental effects on aquatic ecology. Land disposal can be either by dumping in a landfill or by spreading the sludge out over land and allowing natural biodegradation to assimilate the sludge back into the environment. Because of environmental and cost considerations, only incineration and land disposal are presently encountered.

Incineration is actually not a method of disposal at all, but rather a sludge treatment step in which the organics are converted to H_2O and CO_2, and the inorganics drop out as a nonputrescible residue.

The second method of disposal—land spreading—is a science in its infancy. It has become popular only in the last few years, and design and operating data are only now being developed.

The ability of land to absorb sludge and to assimilate it depends on such variables as soil type, vegetation, rainfall, and slope. In addition, the important characteristics of the sludge itself will influence the capacity of a soil to assimilate sludge.

Generally, sandy soils with lush vegetation, low rainfall, and gentle slopes have proven most successful. Mixed digested sludges are spread from tank trucks, and activated sludges are sprayed from both fixed and moving nozzles. The application rate is variable but 100 dry tons/acre/yr is not an unreasonable estimate. Most unsuccessful land application systems can be traced to overloading the soil. Given enough time (and absence of toxic materials) any soil will assimilate sprayed liquid sludge.

There has been some successful use of land application of sludge for fertilization particularly in silviculture operations. Forests and tree nurseries are far enough from population centers to minimize aesthetic objections, and the variable nature of sludge is not so problematic in silviculture as it is in other agricultural applications. Sludge can also be treated and packaged as fertilizer and plant food. The city of Milwaukee has pioneered the drying, disinfection, and deodorizing of sludge, which is packaged and marketed as the fertilizer Milorganite.

Transporting liquid sludge is often expensive, and volume reduction by dewatering is necessary. The solid sludge can then be deposited on land and disked in. A higher rate (tons/acre/yr) can be achieved by trenching where 1-m² (9-ft²) trenches are dug with a backhoe and the sludge deposited and covered. The sludge seems to assimilate rapidly with minimal leaching of nitrates or toxins.

In the last few years a method of chemically bonding the sludge solids so that the mixture "sets" in a few days has found use in industries that have especially critical sludge problems. Although *chemical fixation* is expensive, it is often the only alternative for besieged industrial plants. The leaching from the solid seems to be minimal.

The toxicity can be interpreted in several ways: toxicity to vegetation, toxicity to the animals (including people) who eat the vegetation, and the poisoning of groundwater supplies. Most domestic sludges do not contain sufficient toxins, such as heavy metals, to cause harm to vegetation. The total body burden of heavy metals is of some concern, however. It is possible to precipitate out the metals during sludge treatment, but the most effective means of controlling such toxicity seems to be to prevent these metals from entering the sewage system. Strong, enforced sewer ordinances are necessary and can be cost effective. The regulatory aspects of sludge disposal are discussed further in Chapter 10.

CONCLUSION

Sludge disposal still represents a major headache for many municipalities. Handling and treatment of sludge, particularly when it is concentrated, can pose hitherto unrecognized hazards to the workers who perform these processes. Relief does not seem to be in sight. After all, sludge represents the true residue of our civilization, and its composition reflects our style of living, technological development, and ethical concerns. "Pouring things down the drain" that might cause damage or health problems in the ultimate disposal of the sludge is too often done without thought or malice. We return to the question of moral responsibilities in the chapter on environmental ethics.

PROBLEMS

8.1 A 1-liter cylinder is used to measure the settleability of 0.5 percent suspended solids sludge. After 30 min, the settled sludge solids occupy 600 mL. Calculate the solids concentration of the settled sludge.

8.2 What measures of "stability" would you need if a sludge from a wastewater treatment plant were
   a. placed on the White House lawn
   b. dumped into a trout stream
   c. sprayed on a playground
   d. spread on a vegetable garden

8.3 A sludge is thickened from 2000 mg/L to 17,000 mg/L. What is the reduction in volume, in percent?

Chapter 9
Nonpoint Source Water Pollution

As rain falls and strikes the ground, a complex runoff process begins, and nonpoint source water pollution is the unavoidable result. Even before people entered the picture, the rains came, raindrops picked up soil particles, muddy streams formed, and major water courses became clogged with sediment. Witness the formation of the Mississippi River delta, which has been forming for tens of thousands of years. We can safely surmise that, even before human beings, rivers were "polluted" by this natural series of events; sediment clogged fish gills and fish probably even died. This natural runoff is classified as "background" nonpoint source runoff and not generally labeled as "pollution."

Now view the world as it has been since the dawn of mankind—a busy place where people's activities continue to influence our environment. For years these activities have included irrigation, farming, harvesting trees, constructing buildings and roadways, mining, and disposing of liquid and solid wastes. Each activity has led to disruptions in the surface of the earth's soil and/or involves the application of chemicals to the soil. This increased transport of soil particles, that is, increased sediment loads to watercourses and the application of chemicals to the soil, are generally labeled as pollution.

This "pollution" is the focus of Chapter 9 where we address the runoff process and control technologies applicable to nonpoint source pollution. The focus of this chapter is the six major activities of concern: agriculture, urban stormwater, construction, silviculture, residuals management, and onsite sewage disposal.

135

THE RUNOFF PROCESS

The complex runoff process includes both the detachment and transport of soil particles and leaching and transport chemical pollutants. For the remainder of this chapter, we include all people-induced soil erosion and chemical applications under the term ''nonpoint source pollutant.'' Chemicals can be bound to soil particles and/or be soluble in rainwater; in either case, water movement is the prime mode of transport for solid and chemical pollutants. The *characteristics of the rain* indicate the ability of the rainwater to splash and detach the pollutants. This rain energy is defined by droplet size, velocity of fall, and intensity characteristics of the particular storm.

Soil characteristics impact both the detachment and transport processes. Pollutant detachment is a function of an ill-defined motion of soil stability. Size, shape, composition, and strength of soil aggregates and soil clods all act to determine how readily the pollutants are detached from the soil to begin their movement to streams and lakes within a region. Pollutant transport is influenced by the permeability of the soil to the water. *Soil permeability,* the ease by which water passes through the soil, helps determine the infiltration capabilities and drainage characteristics of the surface receiving the rainfall. Pollutant transport is also a function of soil porosity, which affects storage and movement of water, and soil surface roughness, which tends to create a potential for temporary and long-term detention of the pollutants.

Slope factors also help define the transport component of the nonpoint source problem. The slope gradient as well as slope length influence the flow and velocity of runoff, which in turn influence the quantity of pollutants that are moved from the soil to the water course.

Land cover conditions also affect the detachment and transport of pollutants. Vegetative cover helps to:

- provide protection from the impact of raindrops thus reducing detachment
- make the soil aggregates less susceptible to detachment by protecting soil from evaporation thus keeping the soil moist
- furnish roots, stems, and dead leaves, which help slow overland flow and hold pollutant particles in place

Only a portion of the pollution detached and transported from upland regions in a watershed is actually carried all of the way to a stream or a lake. In many cases, significant portions of the materials are deposited at the base of slopes or on flood plains. The portion of the pollution detached, transported, and actually delivered from its source to the receiving waterway is defined as the *delivery ratio*.

Numerous factors influence the pollutant delivery ratio. Where chemical pollutants are involved, the whole spectrum of factors that determine reaction rates act to limit the delivery ratio: temperature, times of transport, presence of other chemicals, and presence of sunlight, to name just a few. Whenever sediment and/or chemical pollutants become a problem, the list of physical factors becomes quite long and includes:

1. *Magnitude of sediment sources.* Whenever the quantity of sediment available for transport is greater than the capability of the runoff transport system, disposition will occur, and the delivery ratio will be decreased.
2. *Proximity of pollutant sources to receiving waterways.* Pollutants entrapped in runoff often move only short distances but, due to factors such as surface roughness and slope, may be deposited far from the lake or stream. Areas close to a receiving waterway or areas where channel-type erosion takes place may be characterized with a relatively high delivery ratio.
3. *Velocity and volume of water.* The characteristics of the pollutant transport system, particularly the velocity and volume of water from a given storm, impact the delivery ratio. A small storm may not supply enough water to carry a load of pollution to a lake or stream, resulting in a zero or very low delivery ratio. A large, lengthy rainfall may have the opposite effect and transport a very large portion of the pollution that is detached from its source to the receiving waterway.

By understanding rainfall characteristics, soil properties, slope factors, and vegetative covers, the loads of different nonpoint source pollutants to lakes and rivers can be predicted and possibly controlled.

CONTROL TECHNOLOGIES APPLICABLE TO NONPOINT SOURCE POLLUTION

The importance of controlling nonpoint source pollution and the obstacles to achieving that goal are becoming more and more evident. Control over municipal and industrial point sources of pollution historically has received considerable federal and corporate attention through the nationwide construction grants and permit programs. Yet public and private investment to significantly reduce point source pollution may be ill spent in cases where water quality is governed instead by nonpoint source discharges.

Agriculture

Many control measures exist for reducing nonpoint source pollution from agricultural areas. These options range from the management of surface cover and tillage to mechanical conservation measures. Alternative systems have been developed to reduce the erosion potential from tilled (inverted or plowed) land.

Terracing, which breaks the slope of a field into shorter segments, is often applied in fields where contouring and other tillage systems do not offer adequate soil stabilization. Terraces consist of a combination of ridges and channels constructed across the slope that collect surface water from above. Water is either held until it is absorbed or diverted in a controlled manner.

Diversions are large, individually designed terraces constructed across the slope of a field to intercept and divert runoff to a stable outlet. Usually constructed above croplands or above such critical erosion areas as gullies, they act to reduce the volume of runoff entering the problem area.

Other options available to control agricultural nonpoint source pollution include grassed waterways (the use of year-round grasses and areas of surface water movement), and cover plants (grasses, trees, and shrubs planted in critical areas to control severe erosion problems near stream banks).

Silviculture

Silviculture is defined as that part of forest management that deals with the process of utilizing forest crops. Surface water pollution results mainly from three silvicultural activities: the building and use of transportation systems; the harvesting of timber; and intermediate practices such as thinning and spraying for the control of fire, insects, and disease.

From a water quality standpoint, the most critical decisions are made during the planning phase of the crop utilization process. Transportation networks, the most significant nonpoint source of water pollution from silviculture in the Northwest and other areas of the United States, can be designed to minimize pollution runoff. Design considerations include optimizing the number of roadways and their layouts and minimizing the number of times they cross natural water courses. The three surfaces associated with a given segment of roadway offer distinct control options: the cut-slope, the roadway surface itself, and the fill-slope. Geotechnical investigations should precede the construction of any road or staging area. In areas with a slope of greater than 60 percent, it is generally prescribed that roads be built only as a last resort; such log removal techniques as skylining, helicoptering, and ballooning should be investigated. The techniques by which logs are cut and transported influence the amount of nonpoint source pollution.

A component of the runoff problem can be controlled if transportation and harvesting techniques are planned to avoid or minimize the impact on environmentally sensitive flood plains, and buffer strips along waterways should be protected. In addition, chemical applications should follow best management practices. Erosion from staging areas can be controlled by diverting runoff around the site. Bars and ditches can be 70 percent effective in reducing sediment loading to receiving waterways.

Construction

The erosion of soil can cause construction problems onsite as well as water quality problems offsite. In addition, the loss of soil is often regarded as the loss of a valuable natural resource. Home buyers expect a landscaped yard, and lost topsoil is often costly for the contractor to replace. The builder of homes, highways, and other construction views soil erosion as a process that must be controlled to maximize economic gain.

The planning phase of construction activities considers controlled clearing of the proposed construction site as the area to be disturbed during the construction and site restoration phases is held to a minimum. Environmentally sensitive areas must be designated, and if any clearing is required in such areas, it should be limited as much as possible. Such critical areas include: steep slopes, unaggregated soils (sands, and so forth), natural sediment ponds, natural waterways (including intermittent streams), and flood plains.

The planning phase should also consider a comprehensive erosion control system. The components of such a system include planned access as well as techniques for use in the operational and site restoration phases of the construction activity.

During construction, several pollution-abatement techniques appear to be effective. Velocity regulation methods attempt to reduce the rate at which water moves over the construction site. Velocity reduction minimizes particle uptake by the water and can lead to particle disposition in instances where the reduction is sufficient. The result is a decrease in erosion. There are alternative methods of achieving velocity reduction, and all involve the application of some material to the exposed soil: filter inlets, jute mesh, seeding, fertilizing, and mulching can reduce sediment runoff. The best method and its corresponding cost must be determined on a site by site basis.

Stormwater deflection methods attempt to reduce the amount of water passing over the construction site by diverting it before it is properly installed. The dike, however, does not limit or control runoff generated by rain falling directly on the site, and its effectiveness is thus limited.

Stormwater channeling methods attempt to reduce the effect of water passing over the construction site by controlling its movement through the site. Chutes, flumes, and flexible downdrains are effective in certain areas, but their costs are quite high. Their outfall must be handled to ensure that a secondary source of pollution is not created.

Restoration of the site after construction is necessary if water pollution is to be controlled. Regrading costs are necessary expenditures at most construction sites if an effective revegetation process is to be undertaken. Regrading is only effective if it is followed by seeding, fertilizing, and mulching. Such revegetation costs depend on the type of mulch used, and the type of mulch applicable on a given site is highly dependent on the slope of the site. More expensive practices generally are required on steeper slopes. Wood fiber mulch applied by a hydroseeder is very effective on relatively flat landscapes. The more expensive and more permanent excelsior mats and jute netting are also available.

Mining

Mining is discussed in this section under two major headings: surface mining and subsurface mining. The major pollutants from both types of operations are sediments, toxic substances, and acids.

There are several forms of surface mining including strip, open pit, and dredging. Strip mining requires removing a large amount of overlying material (overburden) to expose the desired ore. The distinction between strip mining and open pit mining is that open pit is used in areas where there is relatively little overburden. In such areas, most of the material being removed is the desired mineral, while most of the material being removed during strip mining is overburden and waste. Most strip mining is done to obtain coal, while open pit mining is performed for a range of minerals. Dredging is used to recover minerals from underwater mines with gravel accounting for the majority of dredging production. Hydraulic mining is carried out by directing a high-velocity stream of water at a disperse mineral deposit. It is used almost exclusively for gold.

These forms of mining result in varying amounts of nonpoint source water pollution. Siltation is a major problem, and chemical pollution may occur when surface mining results in accelerating pollution-forming chemical reactions. Water pollution control is generally achieved by changing the conditions responsible for the pollution. Combinations of several techniques are usually required to achieve adequate water pollution abatement. As an example, regrading should be accompanied by revegetation and possibly by water diversion.

Subsurface mines are identified as nonpoint sources of water pollution because the water that runs out of mine openings is often extremely acidic (with a pH as low as 2) and high in metals that may be toxic to aquatic and human life. The source of the water in the mines is usually ground-water that seeps through the porous structure, boreholes, and fractures in the strata adjacent to the mine. The pollutants carried in the water are a result of ores that have been oxidized due to exposure to air. Therefore pollution controls for subsurface mines are aimed at restricting waterflows from the mines and/or maintaining anaerobic atmosphere in the mine.

Urban Stormwater Runoff

The methods for controlling urban stormwater runoff range from non-structural urban housekeeping practices, such as litter control regulations, to structural collection and treatment systems such as settling tanks and possibly even secondary treatment. No single control can be used in all situations or locations. Factors affecting the choice of controls for a given site include: the type of sewerage system (separate or combined), the status of development in the area (planned, developing, or established urban area), and the land use (residential, commercial, or industrial). Controls are grouped into three categories: planning controls, pollutant accumulation controls, and collection and treatment controls. These control categories correspond to those applicable to planned, developing, and developed urban areas.

Several planning controls can be undertaken to reduce the number of pollution sources. Street litter can be reduced by passage and enforcement of antilittering laws. Air pollution, which becomes a source of water pollution when it settles on urban surfaces (particularly streets and roof-tops), can be reduced via effective air pollution abatement planning. Transportation residues such as oil, gas, and grease from cars and partic-ulates from deteriorating road surfaces can be reduced via transportation planning, selection of road surfaces that are less susceptible to deterio-ration, and automobile inspection programs. Preventive actions can be taken as part of land use planning strategies to reduce potential runoff pollution such as avoiding development in environmentally sensitive areas or in areas where urban runoff is an existing problem. Floodplain zoning, one type of land use regulation, often creates a buffer strip that is effective in reducing urban runoff pollution by filtering solids from overland flows and by stabilizing the soils of the floodplain.

Several control techniques prevent pollutants from accumulating on urban surfaces such as streets and parking lots. By preventing such build-up, the total pollutant loadings and the concentration of pollutants in the first flush of an area are reduced. The high concentration of pollutants in this first flush is the cause of many negative water-quality impacts associated with urban runoff. The most common methods of street cleaning include sweeping, vacuuming, and flushing. Street sweeping is the oldest technique and is used in most urban areas. It is also the least expensive of the three. Street sweeping will reduce soil loadings in the runoff but fails to pick up the finer particulates, which often are the more significant source of pollution (biodegradables, toxic substances, nutrients, and so forth). Street vacuuming is more efficient in collecting the small particulates but is more expensive.

Catch basin cleaning refers to the periodic removal of refuse and other solids from catch basins. Significant reductions in biodegradables, nutrients, and other pollutants can result from regular cleaning. The basins can be cleaned by hand or by vacuum.

Urban stormwater runoff pollution can also be controlled after it enters the stormwater drainage system. Detention systems reduce runoff pollutant loadings by retarding the rate of runoff and by encouraging the settling of suspended solids. These systems range from low technology controls such as rooftop storage to intermediate technology controls such as small detention tanks interspersed in the collection network. In general the size and number of units are directly proportional to the effectiveness of the system. Detention basins act like settling tanks and can be expected to remove 30 percent of the biodegradables and 50 percent of the suspended solids in the stormwater.

In storage and treatment systems the first flush of an area is retained in the collection network, in a storage unit, or in a flow equalization basin. The stormwater is then treated at a nearby wastewater treatment facility when the sanitary flow volume and the design capacity of the facility allow. Several innovative storage methods exist, ranging from storing the stormwater in the drainage network through routing the flow using a computerized network of dams and regulators (as in Seattle) to the digging of a subterranean storage tunnel (as in Chicago). The effectiveness of these systems depends on the quantity of pollutants captured, the sizing of the storage units, and the extent of treatment received at the local wastewater facility. The storage capacity is the most critical factor. A debate exists on what size storm should be used as the design storm. The larger the design storm the more costly the system and (generally) the more effective the system. The discussion focuses on whether treatment should be planned for a one-month, a six-month, or even a less common storm.

CONCLUSION

Nonpoint sources contribute major pollutant loadings to the waterways throughout the nation. Control techniques are readily available but vary considerably in both cost and effectiveness, and are typically implemented through a generally confused institutional framework. This institutional setting sometimes poses an obstacle to nonpoint source abatement.

In addition, costs of control for abatement of nonpoint source pollution vary widely and are site specific. Planning controls (preventive measures), however, are relatively inexpensive when compared to operational and site restoration controls (remedial measures) and thus may be the most efficient way to control nonpoint pollution.

PROBLEMS

9.1 Onsite wastewater treatment and disposal systems are also criticized for creating nonpoint source water pollution. Describe the pollutants that are factors in backyards utilizing such systems and identify alternative technologies to control the problems.

9.2 Discuss the wastewater treatment technologies in Chapter 7 that are particularly applicable to help control pollution from urban stormwater runoff.

9.3 Calculate the cost for controlling runoff from the construction of a five-mile highway link across rolling countryside near your home town. Assume hay bales are sufficient along the construction site if they are coupled with burlap barriers located at key locations. No major collection and treatment works are required. Document your assumptions, including labor and materials charges as well as time commitments.

Chapter 10

Water Pollution Law

A complex system of laws requires industries and towns to treat their wastewater flows prior to discharge to receiving waterways. In this system, *common law* and *statutory law* are intertwined to form the regulatory basis for pollution control.

The American legal tradition is based on common law, a body of law different from statutory law as written by Congress and state governments across the nation. This common law is the aggregate body of decisions made in courtrooms as judges decide individual cases. An individual or group of individuals damaged by water pollution or any other wrong (the plaintiffs) historically could seek relief in the courtroom in the form of an injunction to stop the polluter (the defendant) and/or in the form of payment for damages. Because of the possibility that the polluter would have to pay damages, the pollution of waterways can be reduced.

Court rulings in these cases were, and are, based on *precedents*. The underlying theory of precedents is that if a similar case or cases were brought before any court in the past, the present-day judge would violate the rules of fair play if the present-day case were not decided in the same manner and for the same party that the precedent cases dictated. Similar cases, defined to be so by the judge, theoretically have similar endings. If no precedent exists, the plaintiff essentially rolls the dice in hopes of convincing the court to make a favorable precedent-setting decision.

Statutory law, on the other hand, is a set of rules mandated by a representative governing body, be it the Congress of the United States or the state legislatures. Such legislation supplements or changes the effect of existing common law in areas where Congress or state legislatures perceive shortcomings. For example, environmental quality in general and public health in particular were continually harmed under the common laws as they related to dirty water. Common law courts were taking years to reflect changed societal conditions because the courts were bound to precedents set during times in the nation's history when clean water was plentiful and essentially free. Finally, Congress decided to take the initiative with a series of laws aimed at abating water pollution and cleaning the surface waters of the nation.

In this chapter, we discuss the evolution of water law from the common-law courtrooms through the legislative chambers of Congress to the administrative offices of the U.S. Environmental Protection Agency (EPA) and state agencies.

COMMON LAW

To date, common law has concerned itself with the disposition of surface water rather than ground water. There are two major theories of common law as it applies to water. One theory, labeled the *riparian doctrine,* says that conflicts between plaintiffs and defendants must be decided by the ownership of the land underlying or adjoining a body of surface water. The second theory, known as the *prior appropriations doctrine,* takes a different focus and simply states that water use is rationed on a first-come first-serve basis regardless of land ownership. Note that in both these doctrines, the focus is on water quantity, on deciding how to apportion a finite body of clean surface water. Common law is generally unclear about water quality consideration.

The principle underlying the riparian doctrine is that water is owned by the owner of the land underlying or adjoining the stream and that the owner generally is entitled to use the water as long as the quantity is not depleted nor the quality degraded. Riparian land is land bordering a surface waterway.

The doctrine has a somewhat confused history. It was originally introduced to the New World by the French and adopted by several colonies. The English court system eventually adopted the theory as common law in several court cases, and it thus eventually became the official law of the New World. In colonial courtrooms, the riparian doctrine was a workable concept. The land owner was entitled to use the water for domestic purposes such as washing and watering stock, but common law held that it could not be sold to nonriparian parties simply because the water in the stream or river would be diminished in quantity and the downstream user would then not have access to the total flow. Even at the present time in sparsely populated farming areas, where water is plentiful, this system is still applicable.

In urban and more densely populated areas, courts generally found that the riparian doctrine could not be applied in its pure form. Accordingly, several variations or ground rules were developed in the courts: the *principle of reasonable use* and the *concept of prescriptive rights.*

The *principle of reasonable use* holds that a riparian owner is entitled to make reasonable use of the water, taking into account the needs of other riparians. Reasonable use is defined on a case by case basis by the courts. Obviously, this opens tremendous loopholes, which have been used in numerous litigations. Possibly the most famous example is the case of *New York City vs. the States of Pennsylvania and New Jersey.* In the 1920s, New York City began to pipe drinking water from the upper reaches of the Delaware River, and as the city grew, the demand increased until the people downstream from these impoundments found that their rivers had disappeared. Many resort owners simply went out of business. After prolonged court battles, it was finally determined that, since the city did own the land around the impounded streams and the use of this water was "reasonable," the city could continue to use the water. There was some monetary compensation for the downstream riparian owners, but in retrospect the failure of common law is quite evident.

The *concept of prescriptive rights* evolved to the point where, if a riparian owner does not use the water and an upstream user "openly and notoriously" abuses the water quantity or quality, the upstream user is entitled to continue this practice. This concept holds that, through lack of use, the downstream riparian has forfeited the water rights.

This concept was established in a famous case in 1886: *Pennsylvania Coal Co. vs. Sanderson.* Anthracite coal mines north of Scranton, Pennsylvania, at the headwaters of the Lackawanna River, were polluting the river and eventually made it unfit for aquatic life and human consumption. Mrs. Sanderson, a riparian landowner, built a house near the river before the polluted water quality conditions became noticeable. Her intent was to live there indefinitely, but the water quality soon deteriorated, eliminating her opportunities to benefit from the resource. She took the mining company into a courtroom—and lost. The court basically held that the use of the river as a sewer was "reasonable" since the company had been in operation before Mrs. Sanderson had built her house and that, since water pollution was a necessary result of coal mining, the coal company could continue its open and notorious practice. This illustrates another example of where common water law started to break down in its ability to serve the people.

Although historically important, the riparian doctrine is declining in use. It is, after all, applicable to sparsely populated areas with no severe water supply and quality problems. Most of its applicability is limited to areas east of the Mississippi River where there is sufficient rainfall to enable the system to work.

The other important water law concept is the *appropriations doctrine,* which states that water users "first in time" are necessarily "first in line." In other words, if one user put surface water to some "beneficial use" before the next person, the first user is guaranteed that quantity of water for as long as the use demands. Land ownership and user location, upstream or downstream, are irrelevant.

The doctrine began in the mid-nineteenth century as gold miners and ranchers in the arid western United States sought to stake claims to water in the same manner that they staked mining claims. In arid regions, the amount of water in a lake or stream depends on the winter snow-pack rather than on rainfall. Thus, a farmer or rancher who irrigated had a particular concern that development and water use by an upstream farm or ranch would reduce or eliminate stream flow to which he or she had a prior claim. The Reclamation Act of 1902, which provided cheap irrigation water in order to develop the West, made the need to preserve downstream use rights particularly critical.

Since the flow of most streams is highly variable, it is possible to own a water right and a dry stream bed simultaneously. This conflict is resolved under the appropriation doctrine by prior claim modified by priority of uses. For example, if a user has first claim of 1 million gallons per day (mgd), a second claimant has 3 mgd, and the third has 2 mgd, as long as the river flows at 6 mgd everyone is happy. If the flow drops to 4 mgd, the third claimant is completely out of business and can withdraw water only for drinking, cooking, and bathing: Human personal use of water has the highest priority and supersedes other claims. The amount of water allowed for such personal use is, however, no more than 100 gallons per day. If the third claimant happens to be upstream from the first two claimants, almost all of the 4 mgd of water must be permitted to flow past that claimant's water intake.

The Colorado River basin provides an example of the application of the appropriations doctrine. The doctrine generally worked well in the water-short states of the Colorado basin until after World War II. The postwar growth of the sunbelt cities, Los Angeles in particular, led to total appropriation of some waterways. (In some cases more water was appropriated than flowed in the river!) In order to retain a water right, its owner must put the water to "beneficial use." Thus the dams and diversions constructed by those seeking to exercise their water rights enhanced the evaporation rate of some rivers so that the amount of water available was actually reduced. The Angeles River and many tributaries of the Colorado are almost dry today as a result.

The Colorado River and its tributaries are completely appropriated. In order to set aside water for oil shale and uranium mining in the Colorado basin, the developing corporations had to purchase water rights. The cities located along the front range of the Rocky Mountains—Denver, Laramie, and Colorado Springs—have purchased water rights west of the Rockies and divert water through tunnels under the mountains. Albuquerque, Phoenix, and Tucson have also purchased water rights for urban development and divert water from the Rio Grande drainage and the lower Colorado drainage respectively. The looming water shortage in the Colorado basin has led to suggestions that water be diverted from the Columbia River to the Colorado, or even from the Yukon-Charlie system in northern Canada to the Colorado. In 1968, Congress enacted a national water policy that prohibits these massive diversions; this policy is still in force. Clearly, a doctrine of water conservation is needed as a supplement to the prior appropriations doctrine.

As water supply decreased, the monetary value of water rights, which can be bought and sold, increased, leading to overappropriation and the mining of ground water. Overappropriation and overuse of irrigation water result in concentration of pollutants in that water. The recycling of irrigation water in the Colorado River has resulted in an increase in dissolved solids and salinity so great that at the Mexican border Colorado River water is no longer fit for use in irrigation.

As is the case with the riparian doctrine, the appropriations doctrine says very little about water quality. Under the modified appropriations doctrine, the upstream user who is senior in time generally may pollute. If the downstream user is senior, the court has in the past directed payments for losses. If the cost of cleanup is greater than the downstream benefits, however, courts have often found it "reasonable" to allow the pollution to continue. Under the appropriations doctrine, a downstream owner who is not actually using the water has no claim whatever.

The common law theories of public and private nuisance have been found to have applicability in certain cases. Nuisance, however, has found more application in air pollution control and is discussed in some detail in Chapter 21.

STATUTORY LAW

Citing the shortcoming in common law and continued water pollution problems, Congress and state governments have passed a series of laws designed to clean the surface waters across the nation. Although most states had some laws regulating water quality, it was not until 1965 that a concerted push was made to curb water pollution. In that year the U.S. Congress passed the Water Quality Act that, among other provisions, required each state to submit a list of water quality standards and to classify all streams by these standards.

Most states adopted a system similar to the *ambient water quality stream classifications* shown in Table 10-1. Streams were classified according to their anticipated maximum beneficial use. This allowed some states to classify certain streams as low-quality waterways and others as virgin trout streams. The method of stream classification theoretically forces the states to limit industrial and municipal discharges and prevents a stream from decaying further. As progress in pollution control is made, the classifications of various streams can be improved. Lowering a stream classification is generally not allowed by state and federal regulatory agencies. The 1984 amendments to the Safe Drinking Water Act also designated classes of aquifers by their use as drinking-water sources.

In order to attain the desired water quality, restrictions on wastewater discharges are necessary. Such restrictions, known as *effluent standards,* have been used by various levels of government for many years. For example, an effluent standard for all pulp and paper mills may require that the discharge not exceed 50 mg/L BOD. The total loading (in pounds of BOD per unit time) of the pollution or its effect on a specific stream is thus not considered. Using perfectly ''reasonable'' effluent standards, it is still possible that the effluent from a large mill, although it meets the effluent standards, completely destroys a stream. On the other hand, a small mill on a large river, which may in fact be able to discharge even untreated effluent without producing any appreciable detrimental effect on the water quality, must meet the same effluent standards.

This dilemma can be resolved by devoloping a system where minimum effluent standards are first set for all discharges, and these standards are modified based on the actual effect the discharge would have on the receiving watercourse. For example, the large mill above may have a BOD standard of 50 mg/L, but because of the severe detrimental impact it has on the receiving water quality, thus may be reduced to 5 mg/L BOD. This concept requires that each discharge be considered on an individual basis, a process spelled out in the 1972 Federal Water Pollution Control Act, which also established a nationwide policy of zero discharge by 1985. In plain terms, Congress mandated the EPA to ensure that all waste be removed prior to discharge to a receiving waterway. The control mechanism to achieve a reduction in pollution was the EPA's prohibition of any discharge of pollutants into any public waterway unless authorized by a permit. The permit system, known as the National Pollutant Discharge Elimination System (NPDES), is administered by the EPA with direct permitting power transferred to states able to convince the EPA that their administering agency has the authority and expertise to conduct the program. In Wisconsin, for example, the state government has been granted the authority to administer the Wisconsin Pollutant Discharge Elimination System (WPDES).

Table 10-1. Washington State Stream Classification System

Class	Best Use	Fresh waters				Marine waters			
		Min. DO (mg/L)	Max. Temp. (°C)	pH Range	Coliform per 100 mL	Min. DO (mg/L)	Max. Temp. (°C)	pH Range	Coliform per 100 mL
AA	All, fisheries	9.5	16	6.5-8.5	50	7.0	13	7.0-8.5	14
A	All, fisheries	8.0	18	6.5-8.5	100	6.0	16	7.0-8.5	14
B	No fish spawning	6.5	21	6.5-8.6	200	5.0	19	7.0-8.5	100
C	Fish passage, boating	—	24	6.5-9.0	—	4.0	22	6.5-9.0	200
Lake	All	Natural conditions			50				

In situations where an industry wishes to discharge into a municipal sewage treatment system, the industry must agree to contractual arrangements developed with the local governments to ensure compliance with federal industrial pretreatment requirements. For selected industries, pretreatment guidelines are being developed that require facilities to treat their wastewater flows prior to discharge to municipal sewer systems. These pretreatment rules are discussed in detail later in this chapter.

The 1977 amendments to the Clean Water Act have, in recognition of limits of technology and management of wastewater treatment systems to achieve a zero discharge, proposed that eventually all discharges be treated using "best conventional pollutant control technology," even though this would not be 100 percent removal. In addition, the EPA is responsible now for setting effluent limits to a list of about 100 toxic pollutants, which must be controlled using "best available control technology."

An industrial facility has two choices in the disposal of wastewater:

- discharge to a watercourse—in which case a NPDES permit is required, and the discharge will have to be continually monitored
- discharge to a public sewer

The latter method may be preferable if the local publicly owned treatment works (POTW) have the capacity to handle the industrial discharge and if the discharge contains nothing that will poison the POTW secondary treatment system.

Some industrial discharges can, however, cause severe treatment problems in the POTWs, and tighter restrictions on what industries can and cannot discharge into public sewers have become necessary. This has evolved into what is now known as the *pretreatment* program.

Pretreatment Guidelines

The pretreatment regulations have been developed by the EPA. Under these general regulations, any municipal facility or combination of facilities operated by the same authority with a total design flow greater than 5 mgd and receiving pollutants from industrial users is required to establish a pretreatment program. The EPA regional administrator may require that a municipal facility with a design flow of 5 mgd or less develop a pretreatment program if it is found that the nature or the volume of the industrial effluent disrupts the treatment process, causes violations of effluent limitations, or results in the contamination of municipal sludge.

In addition to these general pretreatment regulations, the EPA is developing specific regulations for the 33 major industries listed in Table 10–2.

Table 10–2.   Pretreatment Industries

Timber processing	Laundries
Leather tanning	Soaps and detergents
Steam electric	Machinery
Petroleum refining	Copper working
Iron and steel	Aluminum
Paving and roofing	Plastics
Nonferrous	Batteries
Paint and ink	Coated coils
Printing	Enamel products
Coal mining	Photographic supplies
Ore mining	Foundries
Organics	Adhesives
Inorganics	Explosives
Plastics and synthetics	Gum and wood
Textiles	Pharmaceuticals
Pulp and paper	Pesticides
Rubber	Electroplating

The regulations for each industry are designed to limit the concentrations of certain pollutants that may be introduced into sewerage systems by the respective industries. The standards require limitations on the discharge of pollutants that are toxic to human beings as well as to aquatic organisms, such as cadmium, lead, chromium, copper, nickel, zinc, and cyanide.

Table 10–3 summarizes pretreatment data for six large cities, and presents estimates of the industrial contribution of cadmium to municipal sewerage systems. These data also display the effectiveness of existing pretreatment guidelines in reducing cadmium. One significant conclusion that can be drawn from Table 10–3 is that we actually know very little

Table 10–3.   Pretreatment of Six Sample Cities

	Industrial Contribution: % of Cd in Influent	Effects of Existing Pretreatment Guidelines	
		% Reduction of Cd in Influent	% Reduction of Cd in Sludge
Los Angeles County	86	30	—
Buffalo	100	—	50
Philadelphia	—	80	—
Chicago	38	21	—
Milwaukee	74	—	—
New York City	53	—	—

about industrial discharges in large cities. This lack of understanding is probably equally true for smaller communities.

Drinking Water Standards

Drinking water standards are equally, if not more, important to public health than stream standards. These standards have a long history. In 1914, faced with the questionable quality of potable water in the towns along their routes, the railroad industry asked the U.S. Public Health Service (USPHS) to suggest standards that would describe a good drinking water. As a result of this problem, the first USPHS Drinking Water Standards were born. There was no law passed to require that all towns abide by these standards, but it was established that interstate transportation would not be allowed to stop at towns that could not provide water of adequate quality. Over the years most water supplies in the United States have not been closely regulated, and the high-quality water provided by municipal systems has been as much the result of the professional pride of the water industry personnel as it has of any governmental restrictions.

Because of a growing concern with the quality of some of the urban water supplies and reports that not all waters are as pure and safe as people have always assumed, the federal government passed the Safe Drinking Water Act in late 1974. This law authorizes the EPA to set minimum national drinking-water standards. The EPA has published some of these standards, which are quite similar to the USPHS Water Standards. Some of these representative numbers are shown in Table 10-4. Potable water standards used to describe these contaminants can be divided into three categories: physical, bacteriological, and chemical.

Physical standards include color, turbidity, and odor, all of which are not dangerous in themselves but could, if present in excessive amounts, drive people to drink other, perhaps less safe, water.

Bacteriological standards are in terms of coliforms, the indicators of pollution by wastes from warm-blooded animals. Tests for pathogens are almost never attempted. The present EPA standard calls for a concentration of coliforms of less than 1 per 100 mL water. This standard is a classical example of how the principle of expediency is used to set standards. Before modern water treatment plants were commonplace, the bacteriological standard stood at 10 coliforms/100 mL. In 1946, this was changed to the present level of 1/100 mL. In reality, with modern methods we can attain about 0.01 coliforms/100 mL. It is, however, not expedient to do this because the extra measure of public health attained by lowering the

Table 10–4.   Selected EPA Drinking Water Standards

Physical
  Turbidity   5 units
  Color   15 units
  Odor   3 (threshold odor)
Bacteriological
  Coliforms   1 coliform/100mL

	Suggested (mg/L)	Maximum (mg/L)
Chemical		
Arsenic	0.01	0.05
Chloride	250	—
Copper	1	—
Cyanide	0.01	0.2
Iron	0.3	—
Phenols	0.001	—
Sulfate	250	—
Zinc	5	—

permissible coliform level would not be worth the price we would have to pay.

Chemical standards include a long list of chemical contaminants beginning with arsenic and ending with zinc. Two classifications exist, the first being a suggested limit, the latter a maximum allowable limit. Arsenic, for example, has a suggested limit of 0.01 mg/L. This concentration has from experience been shown to be a safe level even when ingested over an extended period. The maximum allowable arsenic level is 0.05 mg/L, which is still under the toxic threshold but close enough to create public health concern. On the other hand, some chemicals such as chlorides have no maximum allowable limits since at concentrations above the suggested limits the water becomes unfit to drink on the basis of taste or odor. Radioactive contamination is also limited; the limit is stated in terms of maximum individual close.

At present, the only legislation that directly protects ground-water quality is the Safe Drinking Water Act. Increasing pollution of ground water from landfill leachate and inadequately stabilized waste sites is a matter for public concern. Products of the anaerobic degradation of plastics and other synthetic materials are found in ground water in increasing concentration. Some provisions of the Resource Conservation and Recovery Act (RCRA), particularly the provision prohibiting landfill disposal of organic liquids and pyrophoric substances, also provide ground-water protection.

CONCLUSION

Over the years, the battles for clean water have moved from the court-room through the congressional chambers to the administrative offices of EPA and state departments of natural resources. Permitting systems have replaced inconsistent, one-case-at-a-time judicial proceedings as ambient water quality standards and effluent standards are sought. Tough decisions must be faced in the future as regulations are developed for the control of specific hazardous and detrimental substances.

PROBLEMS

10.1 In your hometown, describe the NPDES reporting requirements for the local wastewater treatment facility. What data are required, how often are summary forms completed, and what agency reviews the data on the forms?

10.2 Health departments often require that chlorine be added to water as it enters municipal distribution systems. Discuss the benefits and risks associated with these rules, and describe alternative ways to ensure pota-ble water at the household tap.

10.3 Many industrial processes are water intensive. That is, in order to produce a product that will sell in the marketplace many gallons of water must flow into the factory. Develop a sample listing of such industries and discuss the legal and administrative problems generally associated with securing this water for new factories. Compare and contrast these prob-lems with respect to the generally wet eastern states and dry western states.

10.4 Federal regulations are designed to achieve "zero discharge" of pollutants from point sources located along surface waterways. Land application of liquid waste is an option often proposed in many sections of the nation. Discuss the advantages and disadvantages of such systems particularly in terms of heavy metal pollutants, and outline possible restric-tions on land where such wastes have been applied.

10.5 Assume you work for the EPA and are assigned to propose a stan-dard for the allowable levels of arsenic for household drinking water. What data would you collect, where would you go to get the data (litera-ture and/or laboratory), and in what professions would you seek experts to help guide you?

LIST OF SYMBOLS

    BOD = biochemical oxygen demand
     EPA = U.S. Environmental Protection Agency
  NPDES = National Pollutant Discharge Elimination System
  POTW = publicly owned (wastewater) treatment works
  USPHS = U.S. Public Health Service
  WPDES = Wisconsin Pollutant Discharge Elimination System

Chapter 11
Solid Waste

In this chapter, solid wastes other than hazardous materials and radioactive wastes are considered. Such solid wastes are often called *municipal solid waste* (MSW) and consist of all the solid and semisolid materials discarded by a community. The fraction of MSW produced by a household is called *refuse*.

Refuse until fairly recently was mostly food waste, but new materials such as plastics, new packages for products such as beer cans, and new products such as garbage grinders have all changed the composition of municipal solid waste. Industry creates about 2000 new products each year all of which eventually find their way into municipal refuse and contribute to individual disposal problems.

The components of refuse are *garbage,* food wastes and other organics; *rubbish,* glass, tin cans, and paper; and *ashes,* still a problem where coal is used for heating homes. Lastly, *trash* refers to such larger items as tree limbs, old appliances, and so forth, which are not normally deposited into garbage cans.

The relationship between solid wastes and human disease is intuitively obvious but difficult to prove. For example, if a flea, sustained by a rat that in turn is sustained by an open dump, transmits murine typhus to a human, the absolute proof of the pathway is to find *the* rat and *the* flea, an obviously impossible task. Nevertheless, we are certain that improper solid waste disposal is a true health hazard for we know that at least 22 human diseases are associated with solid wastes.

The two most important vectors[1] of human disease in regard to solid wastes are rats and flies. The fly is a prolific breeder (70,000 flies can be produced in 1 cubic foot of garbage) and a carrier of many diseases, for

---

[1]Vectors are means by which disease organisms are transmitted. Water, air, and food can all be vectors.

example, bacillary dysentery. Rats not only destroy property and infect by direct bite but are also dangerous as carriers of insects that can also act as vectors. For example, the plagues of the Middle Ages were directly associated with the rat populations.

Ground water and drinking water contamination by leachate from solid waste disposal has been growing over the past two decades. Leachate is formed when rain water collects in landfills and stays in contact with the material long enough to leach out and dissolve some of its chemical and biochemical constituents. Leachate can be a major ground water and surface water contaminant, particularly where there is heavy rainfall and rapid percolation through the soil.

As serious as the public health aspects of solid wastes are, health is seldom of primary concern when municipalities decide on a method of disposal. The overriding criterion is still money! What is the cheapest way to "get rid of" this stuff?

In this chapter, the quantities and composition of this "stuff" are discussed first, followed by a brief introduction to disposal options and the specific problem of litter. In Chapter 12, disposal is discussed further, and Chapter 13 is devoted to the problems and promises of recovering energy and materials from refuse.

QUANTITIES AND CHARACTERISTICS OF MUNICIPAL SOLID WASTE

The quantities of MSW generated in a community can be estimated by one of three techniques:

- input analysis
- secondary data analysis
- output analysis

In the first case, the MSW is estimated based on the products people use. For example, if a community purchases 100,000 steel beer cans per week, it can be expected that MSW (including litter pickup) will include about 100,000 beer cans per week. Unfortunately, this technique is very difficult to use in any but very isolated small communities.

Among the characteristics of refuse that are important for developing engineered solutions to refuse management are:

- moisture
- particle size
- chemical makeup

- density
- composition

The moisture concentration of MSW can vary between 15 and 30 percent water with 20 percent being normal. The moisture is measured by drying a sample at 77°C (170°F) for 24 hours and calculating it as

$$M = \frac{w - d}{w} \times 100$$

where   M = moisture content, percent
         w = initial (wet) weight of sample
         d = final (dry) weight of sample

The particle size distribution, especially important in the recovery of materials and energy from refuse, is discussed in Chapter 13. The chemical composition of refuse, covered in Chapters 12 and 13, also affects the recovery of resources. Refuse density can vary from 60–120 kg/m³ (100–200 lb/yd³) for loose refuse, to 300–400 kg/m³ (500–700 lb/y³) in a packer collection vehicle. The density of baled refuse can be as high as 700 kg/m³ (1200 lb/yd³).

Refuse composition is possibly the most important characteristic affecting its disposal or the recovery of materials and energy from refuse. Composition can vary significantly from one community to the next, and with time in any given community.

Refuse composition is expressed either in terms of "as generated" or "as disposed" since during the disposal process moisture transfer takes place thus changing the weights of the various fractions. Table 11–1 summarizes the average solid waste composition for the United States.

COLLECTION

In the United States and most other countries, solid waste is collected by trucks. In some instances, these are open-bed trucks that carry trash or bagged refuse. The usual vehicle, however, is the packer, a truck that uses hydraulic rams to compact the refuse to reduce its volume and thus is able to carry larger loads (Figure 11–1). Commercial and industrial collections are facilitated by the use of containers that are either emptied into the truck using a hydraulic mechanism or where the entire container is carried by the truck to the disposal site (Figure 11–2).

On the average, of the total cost of solid waste management, a full 80

Table 11-1.   Average Composition of MSW in the United States

Category	As Generated (millions of tons)	As Generated (%)	As Disposed (millions of tons)	As Disposed (%)
Paper	37.2	29.0	44.9	34.9
Glass	13.3	10.4	13.5	10.5
Metal	12.1	9.6	12.6	9.8
Ferrous	10.8	8.6		
Aluminum	0.9	0.7		
Other Nonferrous	0.4	0.3		
Plastics	7.7	6.0	7.7	6.0
Rubber and Leather	0.9	0.7	0.6	0.4
Textiles	2.1	1.6	2.2	1.7
Wood	4.9	3.8	4.9	3.8
Food Waste	22.8	17.8	19.1	14.9
Yard Waste	26.0	20.2	20.0	16.3
Miscellaneous	1.9	1.5	2.0	1.6
Total	128.2	100.0	128.5	100.0

Figure 11-1   Packer truck used for residential refuse collection.

Figure 11-2   Containerized collection system (courtesy Dempster Systems).

percent is spent on collection. The common method of collection is by packer truck with three workers: one driver and two loaders. These workers fill the truck and then drive it to the disposal area. The entire operation is a study in inefficiency. The time spent by the loaders traveling to and from the dump, for example, is pure waste. Accordingly, many new devices and methods have been proposed to cut collection cost. Some that are already in use are discussed below.

Garbage grinders reduce the amount of garbage in refuse. If all homes had garbage grinders, the frequency of collection could be cut in half, since the twice-a-week collection in most communities is necessary only because of the rapid decomposition of the garbage component. Obviously, garbage grinders put an extra load on the wastewater treatment plant, but sewage is relatively dilute and ground garbage can easily be accommodated both in the sewers and in treatment plants.

Pneumatic pipes have been installed in some small communities, mostly in Sweden and Japan. The refuse is ground at the residence and sucked through underground lines. One system in the United States is in Walt Disney World in Florida. Collection stations scattered throughout the park receive the refuse, and pneumatic pipes deliver the waste to a central processing plant. It is not at all unreasonable to expect that pneumatic pipes will be the collection method of the future.

Transfer stations are found in almost all larger communities. A typical system involves several stations scattered around a city to which ordinary collection trucks bring the refuse. The drive to the nearest station is fairly short for each truck, so the workers spend more time collecting and less time traveling. At the transfer station, bulldozers cram the refuse into large cans, which in turn take the material to the ultimate disposal site. In some towns the refuse is baled into convenient desk-sized blocks before disposal in a landfill.

Green cans on wheels are now widely used for the transfer of refuse to the truck. The green cans are pushed curbside by the householders and emptied by means of a hydraulic lift. Communities with curbside recycling programs, like Seattle, Washington, and Rockville, Maryland, are beginning to substitute stacked, color-coded curbside bins for green cans. Not only do these systems save money, but they have also a dramatic effect on the incidence of injuries to solid waste collection personnel who have by far the highest lost-time accident rate of any municipal or industrial workers.

Route optimization can result in significant savings to a city. Several computer programs are available for selecting the least-cost routes and collection frequencies. Such optimization techniques have resulted in increased effectiveness and lower cost of refuse collection.

DISPOSAL OPTIONS

The Romans invented city solid waste dumps, and for about 2000 years, municipal solid waste was not perceived as a problem. As cities grew closer together, suburbs developed where municipal dumps would have been. The use of disposable containers and packaging increased vastly increasing the amount of solid waste. In the twentieth century, the disposal of MSW became a serious issue. In response to the disposal problem, many U.S. cities encouraged home incineration of trash, backyard burning, in order to reduce MSW volume. After World War II, some building codes mandated the installation of garbage grinders in new homes in order to reduce the garbage fraction in refuse. Cities like Miami, which simply had no dump sites, built large municipal MSW incinerators.

Growing concern about air and water pollution eventually resulted in widespread prohibition of backyard burning, even of leaves and grass clippings and deemphasis of municipal incineration. The *sanitary landfill* became the accepted method of disposal because it was considered environmentally sound and reasonably inexpensive. The Resource Conservation and Recovery Act (RCRA) prohibited open dumps after 1980 and limited the material that could be landfilled. Landfilling is discussed in Chapter 12.

Unfortunately, appropriate land is not always available, and even if it is, landfilling is not always successful. The operation of landfills is becoming increasingly expensive, and there is well-placed concern about throwing away materials that might prove useful. As a result, the idea of processing wastes to reclaim materials and energy was born. Options for resource recovery are discussed further in Chapter 13.

LITTER

One of the most ubiquitous environment insults is litter. It is also the easiest to control—at least in theory.

Litter is not only unsightly it is also unhealthy (as a breeding ground for rats and other rodents) and damaging to wildlife. Deer and fish, attracted to aluminum can pop tops, ingest them and die in agony. Plastic sandwich bags are mistaken for jellyfish by tortoises and death results. Litter is, in short, a most uncivilized byproduct of our civilization, and considerable efforts have recently been directed toward curbing this insult.

Public awareness campaigns have been ongoing for many years. The most recent efforts toward volunteer participation have been funded by bottle manufacturers and bottlers, hoping to avert legal restrictions on their products. Unfortunately, people still litter and alternative solutions to the problem are being sought.

One report has recommended that from the purely economic standpoint the best way to reduce litter is to hire more street cleaners and road crews to collect the litter. Common sense has happily prevailed and this solution has not been suggested as the final solution to the litter problem.

The availability of trash cans seems to be an important factor influencing littering. A person will walk or ride only a short distance out of the way to deposit waste into a litter can. Increasing the availability and the frequency of maintenance of litter deposits can have a marked effect on litter. Incidentally, experiments have also determined that the type, construction, and color of the litter deposit can all influence the frequency of volunteer participation.

A much more drastic assault on the litter problem is restrictive legislation. The most famous legislative act is the "Oregon Bottle Law," now law in over a dozen other states as well, which prohibits the use of pop-top cans and discourages nonreturnable glass beverage bottles. The effect of the law on litter has been either negligible or fantastic—depending on whose press releases one believes. It has been reported that the percent of beverage containers collected as litter in the states with bottle laws has dropped by 92 percent. Most people do not mind the mild inconveniences and seem to be quite pleased with the results.

The bottling people argue that it is people, not bottles, that cause litter, and they are right in principle. The trick is to solve the problems by changing the habits of the people without resorting to laws. Thus far all such efforts have been unsuccessful.

CONCLUSION

The solid waste problem has three facets: source, collection, and disposal. We have little chance of coping with this problem unless we begin to attack all three areas. The methods of collection and disposal are discussed in the next chapter. The source of solid waste is perhaps the most difficult of the three to tackle. New concepts in packaging, use of natural resources, and evaluation of planned obsolescence are necessary if progress is to be made. In this respect, however, we are swimming against the economic current. The economies in all political systems are oiled with money, and the worth of objects is in terms of cost in money. It is not unreasonable to speculate that this method may some day need to be changed. We may need to create a "new economy" with regard to productivity, obsolescence, and waste before we can begin to be at peace with our ecosystem and have any hopes of long-term survival.

PROBLEMS

11.1 Walk along a stretch of road and collect the litter into two bags, one for beverage containers only and one for everything else. Calculate the
    a.number of items per mile
    b.number of beverage containers per mile
    c.weight of the litter per mile
    d.weight of the beverage containers per mile
    e.percent of beverage containers by count
    f.percent of beverage containers by weight
If you were working for the bottle manufacturers, how would you support your data, as e or f? Why?

11.2 How would it be possible to tax the withdrawal of natural resources? What effect would this have on the economy?

11.3 What effect will the following have on the composition of MSW: (a) garbage grinders, (b) home compactors, (c) no more returnable beverage containers, and (d) a newspaper strike?

11.4 What effect would the Oregon Bottle Law have on your consumption practices? How would you change your lifestyle?

11.5 Drive along a measured stretch of highway and count the pieces of litter visible from the car. (It is best to do this with a friend who does the counting.) Then walk along the same stretch and pick up the litter, counting the pieces and weighing the full bags. What percent of litter (by weight and by piece) is visible from a car?

11.6 On a map of your campus (or any other convenient map), develop an efficient route for refuse collection, assuming that every blockface must be collected.

11.7 Using a study hall or social lounge as a laboratory, study the prevalence of litter by counting the items in the receptacles versus the items improperly disposed of. Each day (weekdays only) vary the conditions as follows:
    Day 1: Normal conditions (baseline)
    Day 2: Remove all receptacles except one
    Day 3: Add additional receptacles (more than normal)
If possible, run several more experiments with different numbers of receptacles. Plot the percent of properly disposed of material versus number of receptacles. Discuss the implications.

Chapter 12

Solid Waste Disposal

The disposal of solid wastes is defined as placement of the waste so it no longer impacts society. This is achieved either by assimilating the residue so it can no longer be identified in the environment (for example, fly ash from an incinerator) or by hiding the wastes well enough so they cannot be readily found.

Solid waste can also be processed so that some of its components can be recovered, a procedure popularly known as recycling. Before disposal or recycling, however, the waste must be collected. All of these—collection, disposal, and/or recovery—form a part of the total solid waste management system. The collection operation is discussed in Chapter 11, and Chapter 13 is devoted to the recovery of energy and materials from refuse. This chapter covers the disposal of solid wastes.

Refuse can be disposed of either as is or after suitable processing. This processing may be thermal or physical, and is performed only for the purpose of converting refuse to a more readily disposable form and not as a method of energy or materials recovery.

DISPOSAL OF UNPROCESSED REFUSE

The only two realistic options for disposal are in the oceans (or other large bodies of water) and on land. In the United States, the former is presently forbidden by federal law, and it is becoming similarly illegal in most other developed nations. Little else need thus be said of ocean disposal except perhaps that its use was a less than glorious chapter in the annals of public health and environmental engineering.

The place for solid waste disposal on land is called a *dump* in the United States and a *tip* in Great Britain (as in "tipping"). The dump is by far the least expensive means of solid waste disposal and thus was the original method of choice for almost all inland communities. The operation of a dump is simple and involves nothing more than making sure that the trucks empty out at the proper spot. Volume is often reduced by setting the refuse on fire, which prolongs dump life.

Rodents, odor, air pollution, and insects at the dump, however, can become serious public health and aesthetic problems, and an alternative method of refuse disposal is necessary. In the United States, dumps have been rendered obsolete and have been replaced mostly by sanitary landfills.

The term *sanitary landfill* was first used for the method of disposal employed in the burial of waste ammunition and other materials after World War II. The concept of refuse burial had, however, been used by several communities in the midwest, and had proven highly successful.

The sanitary landfill (Figure 12–1) differs markedly from open dumps in that the latter are simply places to dump wastes but sanitary landfills are engineered operations, designed, and operated according to acceptable standards.

Sanitary landfilling involves two principles: compaction of the refuse and placement of a cover. Typically, refuse is unloaded either into a trench or at the surface, compacted with heavy machinery, and covered with compacted soil. Daily cover of the refuse is the single feature that renders a landfill much less of a nuisance and health hazard than is a

Figure 12–1   The sanitary landfill.

dump. This cover is from 6 to 12 in thick, depending on the soil composition, and a final cover at least 2 ft thick is used to close the landfill. After closure, a landfill continues to subside, so permanent structures cannot be built on it. Closed landfills do have potential use as golf courses, playgrounds, winter recreation, or even parks or greenbelts.

Landfills may use either the trench or the area method. In the trench method, a trench is dug (the excavated material is used as a cover) and is gradually filled with compacted refuse and cover until grade level is reached. Trench landfills are limited in size but are less of an eyesore than are area landfills.

In the area method, a site is excavated and refuse and cover are built up on the excavated site according to a predetermined plan. Generally, cover material is dug up from one part for use in another part of the site. The refuse is built into compacted mounds, called *cells,* that are approximately trapezoidal in cross section. A completed layer of cells is called a *lift.* Once a lift is completed, another lift is built on the lower lift. The completed landfill consists of many lifts and can extend well above grade.

The selection of a landfill site is a sticky problem. The engineering aspects include: (1) drainage—rapid runoff will lessen mosquito problems, but proximity to streams or well supplies might result in water pollution; (2) wind—it is preferable that the landfill be downwind from the community; (3) distance from collection; (4) size—a small site with limited capacity is generally not acceptable since the trouble of finding a new site is considerable; and (5) ultimate use—can the area be utilized for public or private use after the operation is complete?

Although daily cover limits disease vectors, a working landfill still has a marked and widespread odor during the working day. Flocks of birds that feed at worked landfills are both a nuisance and a hazard to low-flying aircraft and nearby airports. Trucks carrying refuse to the landfill create noise and air pollution problems. In addition, there are the social and psychological problems of a landfill as a neighbor: No one wants one in the back yard.

In years past, "sanitary landfills" were often indistinguishable from dumps and gained a reputation for being bad neighbors. In recent years as more landfills have been operated properly, it has even been possible to enhance property values with a closed landfill site since such a site must remain open space. Acceptable operation and eventual enhancement of the property are understandably difficult to explain to a community.

The landfill operation is actually a biological method of waste treatment. Municipal refuse deposited as a fill is anything but inert. In the absence of oxygen, anaerobic decomposition steadily degrades the organic material to more stable forms. But this process is very slow. After 25 years, the decomposition can still be going strong.

Table 12–1.   Typical Sanitary Landfill Leachate Composition

Component	Typical Value
BOD_5	20,000 mg/L
COD	30,000 mg/L
Ammonia Nitrogen	500 mg/L
Chloride	2,000 mg/L
Total Iron	500 mg/L
Zinc	50 mg/L
Lead	2 mg/L
pH	6.0

The liquid produced during the decomposition process, as well as the water that has seeped through the groundcover and worked its way out of the refuse, is known as *leachate*. This liquid, although small in volume, is extremely high in pollutional capacity. Table 12–1 shows some typical values of leachate composition.

The effect of leachate on groundwater can be severe. In a number of cases, the leachate has polluted wells around a landfill to the point where they ceased to be a source of potable water. One example of such pollution is shown in the problems encountered with a landfill on Long Island. The town of Islip's Sayville landfill was started in a sand and gravel pit in 1933 and extends from about 20 ft above grade to the water table about 30 ft below grade, covering 17 acres. The site is underlain mostly by coarse sand with streaks of gravelly sand.

The leachate plume at Islip's Sayville landfill extends more than 5000 feet downgradient of the site, 170 feet in depth, and up to 1300 feet in width. About 0.22 square mile and one billion gallons of groundwater have been contaminated. Three residential wells near the disposal site and in the leachate plume were contaminated and had to be abandoned. A laundry, sink fixtures, pipes, and a water heater were among the items damaged as a result of the wells' contamination.

A second by-product of a landfill is gas. Since landfills are anaerobic biological reactors, they produce mostly methane and carbon dioxide.

Landfills go through four distinct stages. The first stage is aerobic and may last from a few days to several months during which time aerobic organisms are active and affect the decomposition. As the organisms use up all available oxygen, however, the landfill enters the second stage where anaerobic decomposition begins but where methane-forming organisms have not yet taken hold and the acid formers cause a buildup of CO_2. This stage may also vary with environmental conditions. The third stage is the anaerobic methane production buildup stage during which the percent of CH_4 progressively increases along with an increase in landfill temperature to about 55 °C (130 °F). The last steady-state condition occurs when the fractions of CO_2 and CH_4 are about equal and microbial activity has stabilized.

The amount of methane produced from a landfill can be estimated using the following empirical relationship:

$$CH_aO_bN_c + \frac{1}{4}(4 - a - 2b + 3c)H_2O \rightarrow \frac{1}{8}(4 - a + 2b + 3c)CO_2$$

$$+ \frac{1}{8}(4 + a - 2b - 3c)CH_4$$

(*Note:* a, b, and c are any numbers defining the organic chemicals.) This equation is useful only if the chemical composition of the waste is known.

Unwanted migration of the gas can be prevented by escape vents installed in the landfills. These vents, called *tiki torches,* are kept lit, and the gas is burned off as it is formed.

Improper venting can lead to dangerous accumulation of methane. In 1986, a dozen homes near the Midway landfill in Seattle were evacuated because potentially explosive quantities of methane had leaked through underground fissures into the basements.

The biological aspects of landfills as well as the structural properties of compacted refuse limit the ultimate uses of landfills. Uneven settling is often a problem, and it is generally suggested that nothing be constructed on a landfill for at least two years after completion. With poor initial compaction, it is not unreasonable to expect 50 percent settling within the first five years. The owners of the motel shown in Figure 12–2 learned this fact the hard way.

Landfills should never be disturbed. Not only will this cause additional structural problems, but also trapped gases can be a hazard. Buildings constructed on landfill sites should have spread footings (large concrete slabs) as foundations although some have been constructed on pilings that extend through the fill and onto rock or other adequately strong material.

The cost of operating a landfill varies from about $10 to $30 per ton of refuse and usually represents the least-cost method of acceptable solid waste disposal.

VOLUME REDUCTION PRIOR TO DISPOSAL

Refuse is a bulky material that does not compact easily and thus the volume requirements in landfills are significant. Where land is expensive, the costs of landfilling can be high. Accordingly, various methods of reducing the volume of refuse to be disposed of have been found to be effective.

Under the right circumstances, *incineration* can be an effective treatment of municipal solid waste. Incineration can reduce the volume of

Figure 12–2   A motel that was built on a landfill and experienced differential settling.

waste by a factor of 10 or 20, and the incinerator ash is usually more stable than is the municipal solid waste (MSW) itself. Disposal of the ash can be problematic since heavy metals and some toxic materials will be concentrated in the ash. Incinerators have high capital costs and operating expenses. Air pollution control has effectively doubled the cost of incineration. Understandably the United States is littered with older incinerators that are too expensive to operate.

A schematic of a typical large incinerator is shown in Figure 12–3. The grapple bucket lifts the refuse from a storage pit and drops it into the charging chute. The stoker (in this case a traveling grate) moves the refuse to the furnace area. Combustion occurs both on the stoker and in the furnace. Air is fed under and over the burning refuse. The walls of the furnace are cooled by pipes filled with water. The flue-gases exit through an electrostatic precipitator for controlling particulates and then up the stack.

Smaller incinerators, known popularly as modular incinerators, have been widely used. These units (Figure 12–4) have two chambers, the first chamber being a "smoldering" pit where the refuse is combusted with limited air. The fly ash carried off is combusted in an afterburner, using oil or natural gas as supplemental fuel. This process produces a clear stack gas and can reduce the volume of refuse by 90 percent. It is, however, expensive and dependent on the availability of oil and natural gas for its

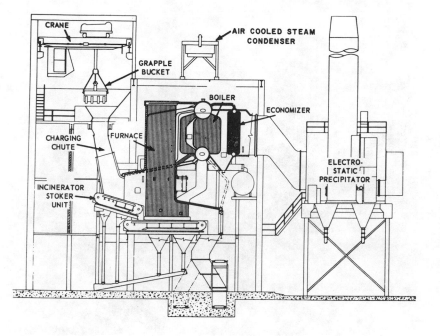

Figure 12-3   Schematic of a typical solid waste incinerator.

operation. Modular incineration in larger facilities is composed of a series of smaller units, each operating independently.

Shredding solid wastes (also known as pulverizing) and then spreading the material on fields has been successful in a number of places. The organics are ground up and are of no interest to rats, and spreading dries the refuse thus avoiding odor and fly problems.

The shredded material does not have to be covered with dirt—a significant advantage over the landfill. Landfill operations are at best difficult during wet or freezing weather, and the capacities of many landfill sites are limited not by available volume but by the dirt available for covering the refuse.

CONCLUSION

At one time, it was easy to dispose of solid waste out of sight and out of the mind of society. In fact, dumping solid waste over the city walls was quite adequate as a method of disposal as civilization developed. In our modern civilization, however, this is no longer possible, and it is

Figure 12–4　Modular incinerators used for solid waste processing (courtesy Consumat).

becoming increasingly difficult to get rid of the stuff so it no longer impacts society.

One potential solution would be to simply redefine solid waste as a resource, and use it for the production of goods for people. This idea is explored in the next chapter.

PROBLEMS

12.1 Suppose the municipal garbage collectors in a town of 10,000 go on strike, and as a gesture to the community, your college or university decides to accept all the city refuse temporarily and pile it on the football field. If all the people did indeed dump the refuse into the stadium, how many days must the garbage collectors be on strike before the stadium is filled to 1 yard deep? (Note: Assume density of refuse as 300 lb/yd, and assume the dimensions of the stadium as 120 yd long and 100 yd wide).

12.2 If a town has a population of 100,000, what is the daily production of wastepaper?

12.3 Describe how you would sell a modular incinerator to your community. Be sure to include the technical, economic, and psychological advantages.

12.4 What would some environmental impacts and effects of depositing dewatered (but sloppy wet) sludge from a wastewater treatment plant into a sanitary landfill be?

12.5 If all of the organic refuse in a landfill were cellulose, $C_6H_{10}O_5$, how much gas (CO_2 and CH_4) would be produced (m^3 of gas/kg refuse)?

12.6 Estimate how much gas you might obtain by capturing the CO_2 and CH_4 from the landfill in your community.

Chapter 13

Resource Recovery

It is becoming increasingly difficult to find new sources of energy and materials to feed our industrial society. Concurrently, we are finding it more and more difficult to locate solid waste disposal sites, and mainly because of transportation requirements, the cost of disposal is escalating exponentially. These two factors—energy and material shortages and fewer and more expensive disposal options—have fueled a new technology called resource recovery.

Except for the process of mass burning of raw refuse, for energy production, the recovery of resources from refuse is primarily a quest for purity. Pure materials can be obtained from mixed municipal solid waste in one of two ways:

1. Separation of the materials is performed by the user, the person who decides to discard the various "consumer products."[1] This is commonly known as *recycling*.
2. Separation is performed after the mixed refuse is collected at a central processing facility. This is commonly referred to as *recovery*.

RECYCLING

There are only two incentives that could be used to convince the public to undertake the separation of refuse components. The first is regulatory, where the governmental agency *dictates* that only separated material will be picked up. Unfortunately, this is not a successful approach in a democracy since public officials advocating unpopular regulations can be removed from office. Recycling in totalitarian regimes, on the other

---

[1]The word *consumer* is clearly a misnomer. If we all were true consumers and not users, there would be no solid waste.

hand, is easy to implement, but this is hardly a compelling argument to abolish democracy.

The second means of achieving cooperation in recycling programs is to appeal to the sense of community spirit and the ethics of environmental concern. Indeed, studies conducted on the feasibility of community waste separation programs, in which householders were asked if they would participate in such programs, attained a 90 to 95 percent positive response. Unfortunately, the *active* response, or the participation in a recycling project, has been much lower. There is a wide gap between what people say they will do (especially if they perceive that the question contains a value component) and how they actually perform. A further complication is that in some inner cities it is difficult to convince people to put refuse in trash cans, much less convince them to separate the refuse into components.

As difficult as recycling is and as small as the fraction of materials removed from the waste stream might be, recycling is still a success in many communities. A number of communities are now achieving 20 percent diversion—mostly newspapers, aluminum cans, and clear glass. The EPA has recently established a national goal of 25 percent recycling, and it appears that many communities will meet this objective. The savings in disposal costs are so significant that several large cities have set much higher, and often unrealistic, goals. Philadelphia, for example, is shooting for 50 percent diversion, an unlikely event.

It is important to recognize that such numbers as "percent diversion" can often be cooked. For example, a community can significantly increase its percent diversion to recycling by simply placing construction rubble into a ravine (for "land reclamation") and count it as recycling.

Community recycling efforts have been successful if:

1. It is easy and convenient for the household to separate the recyclable components.
2. It is efficient and economical for the community to collect the separated material.
3. A community spirit is generated that fosters wide public participation.

The first objective is met by providing recycling containers to the public that are easy to handle and store and are readily identifiable.

Improvements in separated material collection vehicle design over the past few years have been significant. The most effective system seems to be the inclusion of a work station on the truck where the collector can sort the material from the recycling containers while the driver proceeds to the next stop. In this way, undesirable materials can be separated and the product is assured of high purity.

Finally, it is absolutely necessary to establish a sense of community mission or cooperation if a recycling program is to succeed. Public awareness campaigns citing landfill costs and environmental problems, coupled with a sense of "doing the right thing," can coalesce a community into effective concerted action and result in a successful recycling effort.

RECOVERY

Most processes for separation of the various materials in refuse rely on a characteristic or property of the specific material as a code, and this code is used to separate the material from the rest of the mixed refuse. Before such separation can be achieved, however, the material must be in separate and discrete pieces, a condition clearly not met by most components of mixed refuse. A common "tin can," for example, contains steel in its body, zinc on the seam, a paper wrapper on the outside, and perhaps an aluminum top. Other common items in refuse provide equally or even more challenging problems in separation.

One means of assisting in the separation process is to decrease the particle size of refuse, thus increasing the number of particles and achieving a greater number of "clean" particles. This size reduction step, although not strictly materials separation, is commonly a first step in a solid waste processing facility.

Size Reduction

Commonly called *shredding,* the process of size reduction usually consists of a brute-force breakage of particles by swinging hammers in an enclosure. Two types of shredder are used in solid waste processing: the vertical and horizontal hammermills, as shown in Figure 13-1. In the former, the refuse enters the top and must work its way past the rapidly swinging hammers, clearing the space between the hammer tips and the enclosure. Particle size is controlled by adjusting this clearance. In the horizontal hammermill, the hammers swing over a grate that may be changed depending on the size of product required.

Screens

The shredded refuse is often run over screens that separate materials solely by size and do not identify the material by any other property. Consequently, screens are most often used in resource recovery as a classification step prior to a materials separation process. For example, it is possible technically (if not economically) to sort glass into clear and colored

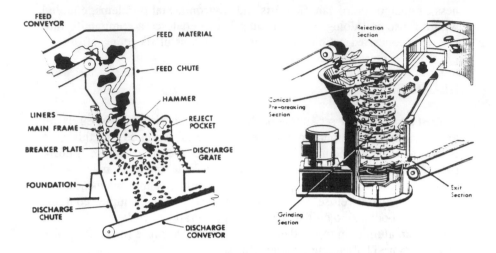

Figure 13-1   Vertical and horizontal hammermills.

fractions by optical coding. This process, however, requires that the glass be of a given size, and screens can be used to produce such a feed to an optical sorter.

Air Classifiers

Materials can be separated according to their aerodynamic properties. In shredded MSW, most of the aerodynamically light materials are organic, and most of the heavy materials are inorganic, thus air classification can produce a refuse-derived fuel (RDF) superior to unclassified shredded refuse. Most air classifiers are similar to the unit pictured in Figure 13-2.

Magnets

Ferrous material is removed using magnets that continually extract the ferrous and reject the remainder. With the belt magnet, recovery of ferrous is enhanced by placing the belt close to the refuse, but this also decreases the purity of the product. A major problem in using belt magnets is the depth of the refuse on the conveyor belt. The heavy ferrous particles tend to settle to the bottom of the refuse carried on a conveyor, and these are then the farthest away from the magnet.

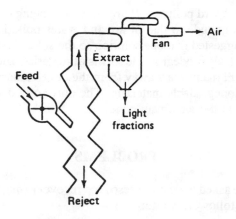

Figure 13-2   Air classifier.

Energy Recovery from the Organic Fraction of MSW

The organic fraction of refuse is, as mentioned earlier, a useful secondary fuel. This shredded and classified product can be used in existing power plants either as a supplemental fuel with coal or fired as the sole fuel in separate boilers. A cross section of a typical heat recovery boiler is shown in Chapter 12. In smaller installations, modular heat recovery boilers, also described in Chapter 12, have found wide use.

Energy Recovery from Unprocessed Refuse (Mass Burn)

Finally, over 100 communities in the United States have opted to install large mass burn units that use unprocessed refuse as fuel. The refuse is burned to produce heat that is in turn used to drive turbines for electrical power production. The ash from such combustion units can cause problems in disposal since they typically contain high levels of heavy metals. Mass burn plants are about twice as expensive as landfills (even after the sale of the electricity) but are often the only alternative for communities that are unable to site new landfills.

CONCLUSION

It is obvious that we must attack the solid waste problem from both ends—from the source as well as from the disposal methods. Solid wastes, often called the "third pollution," are only now being considered a problem equal in magnitude to that of air and water pollution.

It has been suggested that one solution to the solid waste problem is the development of truly biodegradable forms of materials such as plastics and glass. We are still many years away from the development and use of fully recyclable or biodegradable materials. The only truly disposable package available today is the ice cream cone.

PROBLEMS

13.1 You are asked to design a resource recovery (materials separation) system for the following waste:

Component	Fraction by Weight
Newspaper	80
Glass bottles	15
Steel cans	~0
Aluminum cans	~0
Plastics	5
Garbage	~0

Design such a system and draw a schematic diagram. Estimate the quantities you will recover if your plant was constructed in your hometown.

13.2 Aluminum often drives the economics of recycling operations. Estimate the number of aluminum soft-drink and beer cans discarded in your community or at your college. What fraction are recycled? How might this fraction be increased? Offer concrete suggestions.

Chapter 14
Hazardous Waste

For centuries, chemical wastes have been the necessary by-products of developing societies. Here a disposal site, there a disposal site, everywhere a disposal site—all with little or no attention to potential impacts on groundwater quality, runoff to streams and lakes, and skin contact as children played hide and seek in a forest of abandoned 55-gallon drums. Engineering decisions here historically were made by default; little or no handling/processing/disposal planning at the corporate or plant level necessitated quick and dirty decision by mid- and entry-level engineers at the end of production processes. These production engineers solved disposal problems by simply piling or dumping these waste products "out back."

Attitudes began to change in the 1960s and 1970s. As other chapters of this text indicate, air, water, and land were no longer viewed as commodities to be polluted with the problems of cleanup freely passed to neighboring towns or future generations. Individuals responded with court actions against pollution, and governments responded with revised local zoning ordinances, updated public health laws, and new major federal Clean Air and Clean Water Acts. In 1976, the Federal Resource Conservation and Recovery Act (RCRA) was enacted to give the U.S. Environmental Protection Agency specific authority to regulate the generation, transport, and disposal of hazardous waste. Enactment of RCRA has initiated much-needed research into methods of detoxifying or stabilizing hazardous wastes. Engineers are still investigating the most effective ways of managing hazardous waste. This chapter discusses the state of knowledge in the field of hazardous waste engineering, tracing the quantities of wastes generated in the nation from handling and processing options through transportation controls to resource recovery and ultimate disposal alternatives.

MAGNITUDE OF THE PROBLEM

Over the years, the term "hazardous" has evolved in a confusing setting as different groups advocate many criteria for classifying a waste as hazardous. Within the federal government, different agencies used such descriptions as toxic, explosive, and radioactive to label a waste as hazardous.

The federal government attempted to impose a nationwide classification system under the implementation of RCRA, wherein a hazardous waste is defined by the degree of ignitability, corrosivity, reactivity, and/or toxicity. This definition includes acids, toxic chemicals, explosives, and other harmful or potentially harmful waste. In this chapter, this will be the applicable definition of hazardous waste. Radioactive wastes are also hazardous, but because their generation, handling, processing, and disposal differ so drastically from nonradioactive hazards, the radioactive waste problem is addressed separately in Chapter 15.

By this somewhat limited definition, more than 60 million metric tons, by wet weight, of hazardous waste are generated annually throughout the United States. More than 60 percent is generated by chemical and allied products industry, and the machinery, primary metals, paper, and glass products industries each generate between 3 and 10 percent of the nation's total. Approximately 60 percent of the hazardous waste is liquid or sludge. Major generating states, including New Jersey, Illinois, Ohio, California, Pennsylvania, Texas, New York, Michigan, Tennessee, and Indiana, contribute more than 80 percent of the nation's total production of hazardous waste, and most waste is disposed of on the generator's property. More than 80 percent of all disposal is in inadequately designed and operated pits, ponds, landfills, and incinerators.

A hasty reading of these hazardous waste facts points to several interesting, though shocking, conclusions. First, most hazardous waste is generated and inadequately disposed of in the eastern portion of the country. In this region, the climate is wet with patterns of rainfall that permit infiltration and/or runoff to occur. Infiltration permits the transport of hazardous waste into groundwater supplies, and surface runoff leads to the contamination of streams and lakes. Secondly, most hazardous waste is generated and disposed of in areas where people rely on aquifers for drinking water. Major aquifers and well withdrawals underlie areas where the wastes are generated. Thus the hazardous waste problem is compounded by two considerations: The wastes are generated and disposed of in areas where it rains and in areas where people rely on aquifers for supplies of drinking water.

WASTE PROCESSING AND HANDLING

Waste processing and handling are key concerns as a hazardous waste begins its journey from the generator site to a secure, long-term storage facility. Ideally, the waste can be stabilized, detoxified, or somehow rendered harmless in a treatment process similar to those outlined briefly below.

Chemical Stabilization/Fixation. In these processes, chemicals are mixed with waste sludges, the mixture is pumped onto land, and solidification occurs in several days or weeks. The result is a chemical nest that entraps the waste, and pollutants such as heavy metals may be chemically bound in insoluble complexes. Proponents of these processes have argued for building roadways, dams, and bridges using a selected cement as the fixing agent. The environmental adequacy of the processes has not been documented, however, as long-term leaching and defixation potentials are not well understood.

Volume Reduction. Volume reduction is usually achieved in an incineration process. This process takes advantage of the large organic fraction of waste being generated by many industries, but may lead to secondary problems for hazardous waste engineers: air emissions in the stack of the incinerator and ash production in the base of the incinerator. Both by-products of incineration must be addressed in terms of legal, cost, and ethical constraints. Because incineration is often considered a very good method for the ultimate disposal of hazardous waste, we discuss it in some detail later in this chapter.

Waste Segregation. Prior to shipment to a processing or long-term storage facility, wastes are segregated by type and chemical characteristics. Similar wastes are grouped in a 55-gallon drum or group of drums, segregating liquids like acids from solids such as contaminated lab clothing and animal carcasses. Waste segregation is generally practiced to prevent undesirable reactions at disposal sites and may lead to economies of scale in the design of detoxification or resource recovery facilities.

Detoxification. Numerous thermal, chemical, and biological processes are available to detoxify chemical wastes. Options include neutralization, ion exchange, incineration, pyrolysis, aerated lagoons, and waste stabilization ponds. These technologies are extremely waste-specific; ion exchange obviously does not work for every chemical and some forms of heat treatment can be prohibitively expensive for sludges that have a high water content. It is time to call in the chemical engineers whenever detoxification technologies are being considered.

Degradation. Methods exist that chemically degrade some hazardous wastes and render them safer, if not completely safe. Chemical degradation processes, which are very waste-specific, include hydrolysis, to destroy organophosphorus and carbonate pesticides, and chemical dechlorination, to destroy some polychlorinated pesticides. Biological degradation generally involves incorporating the waste into the soil. Landfarming, as it has been termed, relies on healthy soil microorganisms to metabolize the waste components. Such landfarming sites must be strictly controlled for possible water and air pollution that results from overactive or underactive organism populations. For the most part, degradation of hazardous waste is in the research and development stages.

Encapsulation. A wide range of material is available to encapsulate hazardous waste. Ranging from the basic 55-gallon steel drums used throughout the nation, options include concrete, asphalt, and plastics. Several layers of different materials are often recommended such as a steel drum coated with an inch or more of polyurethane foam to prevent rust.

TRANSPORTATION OF HAZARDOUS WASTES

Hazardous wastes are transported across the nation on trucks, rail flatcars, and barges. Transportation of hazardous wastes presents the same hazards and is regulated in the same way as is transportation of other hazardous materials like gasoline. Because many hazardous wastes are often generated in relatively small quantities, truck transportation—often small-truck transportation—is a highly visible and constant threat to public safety and the environment. There are four basic elements in the control strategy for the movement of hazardous waste from a generator.

1. *Haulers.* Major concerns over hazardous waste haulers include operator training, insurance coverage, and special registration of transport vehicles. Handling precautions include gloves, face masks, and coveralls for workers as well as registration of handling equipment to control future use of the equipment and avoid situations where hazardous waste trucks today are used to carry produce to market tomorrow. Schedules for relicensing haulers and checking equipment are part of an overall program for ensuring proper transport of hazardous wastes.

2. *Hazardous Waste Manifest.* The concept of a cradle-to-grave tracking system has long been considered key to proper management of hazardous waste. This "bill of lading" or "trip ticket" ideally accompanies each barrel of waste and describes the content of each barrel to its recipient. Copies of the manifest are submitted to generators and state officials so all parties know each waste has reached its desired destination in a timely manner. This system serves four major purposes: (1) provides the government with a means of tracking waste within a given state and determining

quantities, types, and locations where the waste originates and is ultimately disposed; (2) certifies that wastes being hauled are accurately described to the manager of the processing/disposing facility; (3) provides information for recommended emergency response if a copy of the manifest is not returned to the generator; and (4) provides a data base for future planning within a state. Figure 14–1 illustrates one possible routing of copies of a selected manifest. In this example, the original manifest and five copies are passed from the state regulatory agency to the generator of the waste. Copies accompany each barrel of waste that leaves the generating site and are signed and mailed to the respective locations to indicate the transfer of the waste from one location to another.

3. *Labeling and Placarding.* Before a waste is transported from a generating site, each container is labeled and the transportation vehicle is placarded. Announcements that are appropriate include warnings for explosives, flammable liquids, corrosive material, strong oxidizers, compressed gases, and poisonous and/or toxic substances. Multiple labeling is desirable if, for example, a waste is both explosive and flammable. These labels and placards warn the general public of possible dangers and assist emergency response teams as they react in the event of a spill or accident along a transportation route.

4. *Accident and Incident Reporting.* In transportation regulation, "incidents" involve damage to the waste package without involving the vehicle; "accidents" involve the vehicle carrying the waste as well as the waste itself. Accidents involving hazardous wastes must be reported immediately to state regulatory agencies and local health officials. Accident reports that are submitted immediately and indicate the amount of materials released, the hazards of these materials, and the nature of the failure that caused the accident can be instrumental in containing a waste and cleaning the site. For example, if liquid waste can be contained, groundwater and surface water pollution may be avoided.

RESOURCE RECOVERY ALTERNATIVES

Resource recovery alternatives are based on the premise that one man's waste is another man's prize. What may be a worthless drum of electroplating sludge to the plating engineer may be a silver mine to an engineer skilled in metals recovery. In hazardous waste management, two types of systems exist for transferring this waste to a location where it is viewed as a resource: *hazardous waste materials transfers* and *hazardous waste information clearinghouses*. In practice one organization may display characteristics of both of these pure systems.

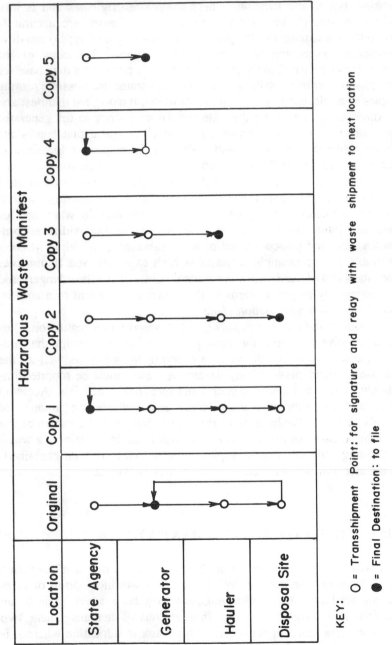

KEY:

○ = Transshipment Point: for signature and relay with waste shipment to next location

● = Final Destination: to file

Figure 14–1   Possible routing of copies of a hazardous waste manifest.

Information Clearinghouses

The pure clearinghouse has limited functions. Basically, these institutions offer a central point for collecting and displaying information about industrial wastes. Their goal is to introduce interested potential trading partners to each other through the use of anonymous advertisements and contacts. Clearinghouses generally do not seek customers, negotiate transfers, set prices, process materials, or provide legal advice to interested parties. One major function of a clearinghouse is to keep all data and transactions confidential so trade secrets are not compromised.

In comparison to the clearinghouse concept, a pure materials exchange has many complex functions. A transfer agent within the exchange typically identifies generators of waste and potential users of the waste. The exchange will buy or accept waste, analyze its chemical and physical properties, identify buyers, reprocess the waste as needed, and sell it at a profit.

Clearinghouses and exchanges have been attempted with some success in the United States. However, a longer track record exists in Europe. Belgium, Switzerland, West Germany, most of the Scandinavian countries, and the United Kingdom all have experienced some success with exchanges. The general characteristics of European waste exchanges include:

- operation by the national industrial associations
- services offered without charge
- waste availability made known through published advertisements
- advertisements discuss chemical and physical properties, as well as quantities, of waste
- advertisements are coded to keep the identity of offerer confidential

Much can be learned in the United States from these experiences.

Five wastes are generally recognized as having transfer value: (1) wastes having high concentrations of metals, (2) solvents, (3) concentrated acids, (4) oils, and (5) combustibles for fuel. That is not to say these wastes are the only transferable items. Transformed from waste to resource in one European exchange were:

- foundry slag, 50-60 percent metallic Al, 400 ton/yr
- methanol, 90 percent with trace mineral acids, 150 m^3/yr
- cherries, deep frozen, 4 tons

One person's waste can truly be another person's valued resource.

HAZARDOUS WASTE MANAGEMENT FACILITIES

Siting Considerations

In selecting a site, all of the relevant "ologies" must be considered: hydrology, climatology, geology, and ecology as well as current land use, environmental health, and transportation. Risk analysis is an important part of the siting process (Chapter 1).

Hydrology. Hazardous waste landfills should be located well above historically high groundwater tables. Care should be taken to ensure that a location has no surface or subsurface connection such as a crack in confining strata, between the site and a watercourse. Hydrological considerations limit direct discharge of wastes into groundwater or surface water supplies.

Climatology. Hazardous waste management facilities should be located outside the paths of recurring severe storms. Hurricanes and tornadoes disrupt the integrity of landfills and incinerators and cause immediate catastrophic effects on the surrounding environment and public health in the region of the facility. In addition, areas of high air pollution potential should be avoided in site selection processes. These areas include valleys where winds and/or inversions act to hold pollutants close to the surface of the earth as well as areas on the windward side of mountain ranges such as areas similar to the Los Angeles area where long-term inversions are prevalent.

Geology. A disposal or processing facility should only be located on stable geologic formations. Impervious rock, which is not littered with cracks and fissures, is an ideal final liner for hazardous waste landfills.

Ecology. The ecological balance must be considered as hazardous waste management facilities are located in a region. Ideal sites in this respect include areas of low fauna and flora density, and efforts should be made to avoid wilderness areas, wildlife refuges, and migration routes. Areas with unique plants and animals, especially endangered species, should also be avoided.

Alternative Land Use. Areas with low ultimate land use should receive prime consideration as facilities are sited in a region. Areas with high recreational use potential should be avoided because of the increased possibility of direct human contact with the wastes.

Environmental Health. Landfills and processing facilities should be located away from private wells, away from municipal water supplies, and away from high population densities. Flood plains should be avoided, at least up to the 100-year storm level.

Transportation. Transportation routes to facilities are a major consideration in siting hazardous waste management facilities. Such facilities should be accessible by all-weather highways to avoid spills and accidents during periods of rain and snowfall. Ideally, the closer a facility is to the generators of the waste, the less likely are spills and accidents as the wastes move along the countryside.

Socioeconomic Factors.   Factors that could make or break an effort to site a hazardous waste management facility fall under this major heading. Such factors, which range from citizen acceptance to long-term care and monitoring of the facility, are:

1.   Citizen acceptance and public education programs: Will local townspeople permit it?
2.   Land use changes and industrial development trends: Does the region wish to experience the industrial growth that is induced by such facilities?
3.   User fee structures and recovery of project costs: Who will pay for the facility, can user changes be used to induce industry to reuse, reduce, or recover the resources in the waste?
4.   Long-term care and monitoring: How will post-closure maintenance be guaranteed and who will pay?

All are critical concerns in a hazardous waste management scheme.

Mixed Waste

The term "mixed waste" refers to mixtures of hazardous and radioactive wastes. (Organic solvents used in liquid scintillation counting are an excellent example.) Siting a new mixed waste facility is virtually impossible at present because in many cases the laws governing handling of chemically hazardous waste conflict with those governing handling of radioactive waste. Existing radioactive waste disposal sites contain some mixed waste.

Incinerators

Incineration is a controlled process that uses combustion to convert a waste to a less bulky, less toxic, or less noxious material. The principal products of incineration from a volume standpoint are carbon dioxide, water, and ash, but the products of primary concern due to their environmental effects are compounds containing sulfur, nitrogen, and halogens. When the gaseous combustion products from an incineration process contain undesirable compounds, a secondary treatment such as afterburning, scrubbing, or filtration is required to lower concentrations to acceptable levels prior to atmospheric release. The solid ash products from the incineration process are a major concern and must reach adequate ultimate disposal. The advantages of incineration as a means of disposal for hazardous waste are

1. Burning wastes and fuels in a controlled manner has been carried on for many years and the basic process technology is available and reasonably well developed. This is not the case for some of the more exotic chemical degradation processes.
2. Incineration is broadly applicable to most organic wastes and can be scaled to handle large volumes of liquid waste.
3. Large expensive land areas are not required.

The disadvantages of incineration include:

1. The equipment tends to be more costly to operate than many other alternatives.
2. It is not always a means of ultimate disposal in that normally an ash remains that may or may not be toxic but that in any case must be disposed of properly.
3. Unless controlled by applications of air pollution control technology, the gaseous and particulate products of combustion can be hazardous to health or damaging to property.

The decision to incinerate a specific waste will therefore depend first on the environmental adequacy of incineration as compared to other alternatives and second on the relative costs of incineration and the environmentally sound alternatives.

The variables that have the greatest effect on the completion of the oxidation of wastes are waste combustibility, dwell time in the combustor, flame temperature, and the turbulence present in the reaction zone of the incinerator. The combustibility is a measure of the ease with which a material can be oxidized in a combustion environment. Materials with a low flammability limit, low flash point, and low ignition and auto ignition temperatures may be combusted in a less severe oxidation environment—that is, at a lower temperature and with less excess oxygen.

Of the three "Ts" of good combustion—time, temperature, and turbulence—only the *temperature* may be readily controlled after the incinerator unit is constructed. This can be done by varying the air-to-fuel ratio. If solid carbonaceous waste is to be burned without smoke, a minimum temperature of 760 °C (1400 °F) must be maintained in the combustion chamber. Upper temperature limits in the incinerator are dictated by the refractory materials available. Above 1300 °C (2400 °F) special refractories are needed.

The degree of *turbulence* of the air for oxidation with the waste fuel will affect the incinerator performance significantly. In general, both mechanical and aerodynamic means are utilized to achieve mixing of the air and fuel. The completeness of combustion and the time required for complete combustion are significantly affected by the amount and the effectiveness of the turbulence.

The third major requirement for good combustion is *time.* Sufficient time must be provided to the combustion process to allow slow-burning particles or droplets to burn completely before they are chilled by contact with cold surfaces or the atmosphere. The amount of time required depends on the temperature, fuel size, and degree of turbulence achieved.

The type and form of waste will dictate the type of combustion unit required. A number of control methods have been successfully developed for applications where the pollutants are in the form of fumes or gas. If the waste gas contains organic materials that are combustible, incineration should be considered as a final method of disposal. When the amount of combustible material in the mixture is below the lower flammable limit, it may be necessary to add small quantities of natural gas or other auxiliary fuel to sustain combustion in the burner. Thus economic considerations are critical in the selection of incinerator systems because of the high costs of these additional fuels.

Boilers for some high-temperature industrial processes can serve as incinerators for toxic or hazardous carbonaceous waste. Cement kilns, which must operate at temperatures in excess of $1400\,°C$ ($2550\,°F$) in order to produce cement clinker, can use organic solvents as fuel, and this provides an acceptable method of waste solvent and waste oil disposal.

Incineration is also a possibility for the destruction of liquid wastes. Liquid wastes may be classified into two types from a combustion standpoint: combustible liquids and partially combustible liquids. Combustible liquids include all materials having suffcient caloric value to support combustion in a conventional combustor or burner. Noncombustible liquids cannot be treated or disposed of by incineration and include materials that would not support combustion without the addition of auxiliary fuel and would have a high percentage of noncombustible constituents such as water.

When starting with a waste in liquid form, it is necessary to supply sufficient heat for vaporization in addition to raising it to its ignition temperature. In order that a waste may be considered combustible, several rules of thumb should be used. The waste should be pumpable at ambient temperature or capable of being pumped after heating to some reasonable temperature level. Since liquids vaporize and react more rapidly when finely divided in the form of a spray, atomizing nozzles are usually employed to inject waste liquids into incineration equipment whenever the viscosity of the waste permits atomization. If the waste cannot be pumped or atomized, it cannot be burned as a liquid but must be handled as a sludge or solid.

In order to support combustion in air without the assistance of an auxiliary fuel, the waste must generally have a caloric value of 18,500–23,000 kJ/kg (8000–10,000 Btu/lb) or higher. Liquid waste having a heating value below 18,500 kJ/kg (8000 Btu/lb) is considered a partially combustible material and requires special treatment.

Incineration of wastes that are not pure liquids but that might be considered sludges or slurries is also an important waste disposal problem. Incinerator types applicable for this kind of waste would be fluidized bed incinerators, rotary kiln incinerators, and multiple hearth incinerators.

Incineration is not a total disposal method for many solids and sludges because most of these materials contain noncombustibles and have residual ash. Complications develop with the wide variety of materials that must be burned. Controlling the proper amount of air to give combustion of both solids and sludges is difficult, and with most currently available incinerator designs, this is impossible.

The types of incinerators applicable to solid wastes are open pit incinerators and closed incinerators such as rotary kilns and multiple hearth incinerators. Generally, the incinerator design does not have to be limited to a single combustible or partially combustible waste. Often it is both economical and feasible to utilize a combustible waste, either liquid or gas, as the heat source for the incineration of a partially combustible waste that may be either liquid or gas.

Experience indicates that wastes, which contain only carbon, hydrogen, and oxygen and which can be handled in power generation systems, can be destroyed in a way that reclaims some of their energy content. These types of wastes may also be judiciously blended with wastes having low energy content such as the highly chlorinated organics in order to minimize the use of purchased fossil fuel. On the other hand, rising energy costs will not be a significant deterrent to the use of thermal destruction methods where they are clearly indicated to be the most desirable method on an environmental basis.

Air emissions from hazardous waste incinerators include the common air pollutants discussed in Chapter 17. In addition, inadequate incineration can result in emission of some of the hazardous materials that the incineration was intended to destroy. Incomplete combustion, particularly at relatively low temperatures, can also result in production of a class of compounds known collectively as *dioxins,* including both polychlorinated dibenzodioxins (PCDD) and polychlorinated dibenzofurans (PCDF). The compound in this class that has been identified as a carcinogen and teratogen is 2,3,7,8-tetrachlorodibenzo-*p*-dioxin (2,3,7,8-TCDD), shown

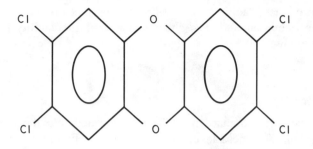

Figure 14–2   2,3,7,8-dichlorodibenzo-*p*-dioxin.

in Figure 14–2. There is growing public concern about TCDD emissions from incinerators.

TCDD was first recognized as an oxidation product of trichlorophenol herbicides (2,4-D and 2,4,5-T, one of the ingredients of Agent Orange)[1] In1977, it was one of the PCDDs found present in municipal incinerator fly ash and air emissions and has subsequently been found to be a consti-tuent of gaseous emissions from virtually all combustion processes, in-cluding trash fires and barbecues. TCDD is degraded by sunlight in the presence of water.

The acute toxicity of TCDD in animals is extremely high (LD_{50} in hamsters of 3.0 mg/kg); carcinogenesis and genetic effects (teratogenesis) have also been observed in chronic exposure to high doses in experimental animals. In humans, the evidence for these adverse effects is mixed. While acute effects like skin rashes and digestive difficulties have been observed on high accidental exposure, these are transitory. Public concern has focused on chronic effects, but existing evidence for either carcinogenesis or birth defects in humans from chronic TCDD exposure is inconsistent. Regulations governing incineration are designed to limit TCDD emission to below measureable quantities; these limits can usually be achieved by the proper combination of temperature and residence time.

Engineers should understand, however, that public concern about TCDD (and dioxins in general) has occasionally reached hysterical pro-portions and is a major factor in opposition to incinerator siting.

Landfills

Landfills must be adequately designed and operated if public health and the environment are to be protected. The general components that go into

---

[1]F. H. Tschirley, *Scientific American* 254 (1986): 29–35.

the design of these facilities, as well as the correct procedure to follow during the operation and post-closure phase of the facility's life, are discussed below.

Design

Three levels of safeguard must be incorporated into the design of a hazardous landfill. These levels are displayed in Figure 14-3. The primary system is an impermeable liner, either clay or synthetic material, coupled with a leachate collection and treatment system. Infiltration can be minimized with a cap of impervious material overlaying the landfill and sloped to permit adequate runoff and to discourage pooling of the water.

The objectives are to prevent rainwater and snow melt from entering the soil and percolating to the waste containers and, in case water does enter the disposal cells, to collect and treat it as quickly as possible. Side slopes of the landfill should be a maximum of 3:1 to reduce stress on the liner material. Research and testing of the range of synthetic liners must be viewed with respect to a liner's strength, compatibility with wastes, costs, and life expectancy. Rubber, asphalt, concrete, and a variety of plastics are available, and such combinations as polyvinyl chloride overlaying clay may prove useful on a site-specific basis.

A leachate collection system must be designed by contours to promote movement of the waste to pumps for extraction to the surface and subsequent treatment. Plastic pipes, or sand and gravel, similar to systems in

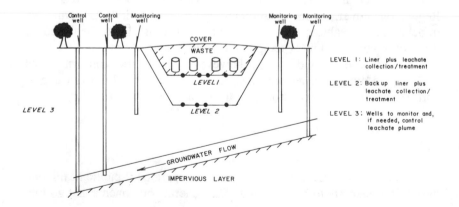

Figure 14-3   Three levels of safeguard in hazardous waste landfills.

municipal landfills and used on golf courses around the country, are adequate to channel the leachate to a pumping station below the landfill. One or more pumps direct the collected leachate to the surface where a wide range of waste-specific treatment technologies are available, including:

- sorbent material: carbon and fly ash arranged in a column through which the leachate is passed
- packaged physical-chemical units, including chemical addition and flash mixing, controlled flocculation, sedimentation, pressure filtration, pH adjustment, and reverse osmosis

The effectiveness of each method is highly waste specific and tests must be conducted on a site-by-site basis before a reliable leachate treatment system can be designed. All methods produce waste sludges that must reach ultimate disposal.

A secondary safeguard system consists of another barrier contoured to provide a backup leachate collection system. In the event of failure of the primary system, the secondary collection system conveys the leachate to a pumping station, which in turn relays the wastewater to the surface for treatment.

A final safeguard system is also advisable. This system consists of a series of discharge wells up-gradient and down-gradient to monitor groundwater quality in the area and to control leachate plumes if the primary and secondary systems fail. Up-gradient wells act to define the background levels of selected chemicals in the groundwater and to serve as a basis for comparing the concentration of these chemicals in the discharge from the down-gradient wells. This system thus provides an alarm mechanism if the primary and secondary systems fail.

If gas generation is possible in a hazardous waste landfill, a gas collection system must be designed into the landfill. Sufficient vent points must be allowed so that the gas generated may be burned off continuously or processed prior to its emission into the atmosphere.

Operation

As waste containers are brought to a landfill site for burial, specific precautions should be taken to ensure the protection of public health, worker safety, and the environment. Wastes should be segregated by physical and chemical characteristics and buried in the same cells of the landfill. Three-dimensional mapping of the site is useful for future mining of these cells for resource recovery purposes. Observation wells with continuous monitoring should be maintained, and regular core soil samples should be taken around the perimeter of the site to verify the integrity of the liner materials.

Site Closure

Once a site is closed and does not accept more waste, the operation and maintenance of the site must continue. The impervious cap on top of the landfill must be inspected and maintained to minimize infiltration. Surface water runoff must be managed, collected, and possibly treated. Continuous monitoring of surface water, groundwater, soil, and air quality is necessary as ballooning and rupture of the cover material may occur if gases produced and/or released from the waste rise to the surface. Waste inventories and burial maps must be maintained for future land use and waste reclamation. A major component of postclosure management is maintaining limited access to the area.

CONCLUSION

Hazardous waste is a relatively new concern of environmental engineers. For years, the necessary by-products of an industrialized society were piled "out back" on land that had little value. As time passed and the rains came and went, the migration of harmful chemicals moved hazardous waste to the front page of the newspaper and into the classroom. Engineers employed in all public and private sectors must now face head-on the processing, transport, and disposal of these wastes.

Hazardous waste is appropriately addressed at the "front end" of the generation process: either in maximizing resource recovery in house, using industry-wide clearinghouses and exchanges, or detoxification at the site of generation. Storage, landfilling in particular, is at best a stopgap measure for hazardous waste handling.

PROBLEMS

14.1 Assume you are an engineer working for a hazardous waste processing firm. Your vice president thinks it would be profitable to locate a new regional facility near the state capitol. Given what you know about that region, rank the factors that distinguish a good site from a bad one. Discuss the reasons for this ranking: Why, for example, are hydrologic considerations more critical in that region than are geological ones?

14.2 If you were a town engineer just informed of a chemical spill on Main Street, sequence your responses. List and describe the actions your town should take for the next 48 hours if the spill is relatively small (100–500 gallons) and is confined to a small plot of land.

14.3 Compare and contrast the design considerations of a hazardous waste landfill with those of a conventional municipal refuse landfill.

14.4 As town engineer, design a system to detect and stop the movement of hazardous wastes into your municipal refuse landfill.

Chapter 15

Radioactive Waste

In the late 1800s, French scientists determined that uranium minerals routinely emitted invisible radiation capable of passing through apparently solid objects. Building on the research, Pierre and Marie Curie were able to isolate two new chemical elements from the uranium minerals: polonium and radium; and these newly discovered elements were observed to produce radiation that was even more intense than uranium emissions. Subsequently, it was determined that the radiation was a result of an atomic disintegration as atoms of the minerals somehow disintegrated spontaneously through time.

This chapter presents a general background discussion of the interaction of ionizing radiation with matter and of the environmental effects of radionuclides. The chapter highlights radioactive waste as a pollutant, discusses the impacts of ionizing radiation on the natural environment and on public health, and summarizes options available to environmental engineers for treating and stabilizing radioactive waste.

RADIATION

The Curies and their contemporaries classified the radiation from uranium minerals (pitchblende) into three types, according to the direction of deflection in a magnetic field. These three types of radiation were called alpha (α), beta (β), and gamma (γ) radiation. Becquerel recognized that gamma radiation was the equivalent of the "x-rays" discovered by Roentgen. In 1932, Chadwick identified the neutron as the highly penetrating radiation that results when beryllium is bombarded with alpha particles. Modern physics has subsequently identified emission of positrons, muons, and pions, but not all are of equal concern to the environmental engineer.

Table 15–1.   Dangerous Radioactive Products

Product	Type of Radiation	Halflife
Krypton-85 (Kr^{85})	beta and gamma	10 years
Strontium-90 (Sr^{90})	beta	20 years
Iodine-131 (I^{131})	beta and gamma	8 days
Cesium-137 (Cs^{137})	beta and gamma	30 years
Tritium (TI)	beta	12 years
Cobalt-60 (Co^{60})	beta and gamma	5 years
Carbon-14 (C^{14})	beta	5770 years

The significant problems associated with the management of radioactive wastes require a basic understanding of alpha, beta, and gamma emissions and emitters and understanding of the effect of neutrons.

Often, the original radionuclide, called the parent, decays to a nucleus that is also unstable, called the daughter; the daughter often decays even further. Chains where radioactive daughters produce radioactive second daughters, which in turn produce radioactive third daughters and on and on and on, are not uncommon. For example, for the decay chain beginning with U^{238} we observe ten steps before the stable Pb^{207} is reached. Radioactive decay is commonly described in terms of the *halflife*, or the time for one half of the radioactive atoms to have disintegrated. The halflives of some of the most significant radioactive isotopes are shown in Table 15–1.

A useful figure of merit for radioactive decay is:

- after 10 halflives, 10^{-3}, or 0.1 percent, of the original quantity of radioactive material is left
- after 20 halflives, 10^{-6}, or 0.0001 percent, of the original quantity of radioactive material is left.

Alpha (α), Beta (β), and Gamma (γ) Radiation

Emissions from radioactive nuclei are characterized, in general, as ionizing radiation, because collision between these emissions and an atom or molecule ionizes that atom or molecule. Alpha, beta, and gamma rays or particles can be characterized further by their movement in an electric or magnetic field. Apparatus for such a characterization is shown in Figure 15–1. A beam of disintegrating radioactive atoms is aimed, with a lead barrel, at a fluorescent screen, which is designed to glow when hit by the radiation. Two alternately charged probes direct the positively charged α radiation and negatively charged β radiation accordingly. The γ radiation is seen to be "invisible light," a stream of neutral particles

Barrel (Pb)

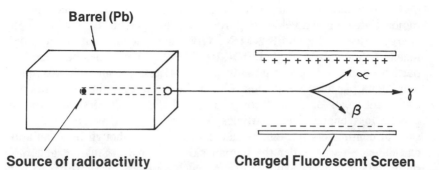

Source of radioactivity   **Charged Fluorescent Screen**

Figure 15-1   Controlled measurement of alpha (α), beta (β), and gamma (γ) radiation.

that passes through the electromagnetic force field. Alpha and beta emissions are typically classified as particules, while gamma emissions consist of electromagnetic radiation (waves).

Alpha radiations, identified as being physically identical to the nuclei of helium atoms, stripped of their planetary electrons with only 2 protons and 2 neutrons remaining, are emitted from the nucleus of selected radioactive atoms with a kinetic energy of somewhere between 4 and 10 MeV.[1] The particles have a mass of about 4 amu (6.642×10^{-4}g) and a positive charge of 2.[2] As these charged particles travel along at approximately 10,000 miles per second, they interact with other atoms. Each interaction transfers a portion of the energy to the electrons of these other atoms and results in the production of an ion pair; that is, a negative electron with an associated positive ion. The huge number of interactions produces between 30,000 and 100,000 of these ion pairs per centimeter of air traveled, resulting in rapid depletion of the alpha energy. Consequently, the range of the alpha particle is only 1–8 cm in air. If such a particle runs into a solid object, notably human skin cells, the energy is rapidly dissipated. Thus alpha particles present no direct problem of external radiation damage to humans but could cause health problems when emitted inside the body where protective layers are not present to diffuse the energy. Even the strongest alpha particles are stopped by the epidermal layer of the skin and rarely reach the sensitive layers. Humans are typically contaminated with material that emits alpha particles only if the material is inhaled, ingested, or absorbed in a skin wound.

---

[1] Mass is designated as energy in the equation $E = mc^2$. One MeV = 10^6 electron volts (ev). One electron volt = 1.603×10^{-12}erg.
[2] This charge, in atomic mass units (amu), is expressed in units relative to an electron with a negative charge of 1.

Beta radiation is electrons that are also emitted from the nucleus of a radioactive atom at a velocity approaching the speed of light with kinetic energy between 0.2 and 3.2 MeV. Given their lower mass of approximately 5.5×10^{-4} amu (9.130×10^{-28}g), the interactions between beta particles and the atoms of pass-through materials are much less frequent than alpha particle interactions. Fewer than 200 ion pairs are formed in each centimeter of flight through air, and the resulting slower rate of energy loss enables beta particles to travel several meters in air and several centimeters through human tissue. Thus although beta radiation can damage tissue under the human skin, internal organs are generally protected. However, exposed organs such as eyes are sensitive to beta damage.

Gamma radiation is identified as invisible, electromagnetic rays emitted from the nucleus of radioactive atoms. These rays are like medical x-rays, in that they are composed of photons. Because of their neutral charge, gamma photons collide randomly with the atoms of the material as they pass through. A typical gamma ray, with an energy of about 0.7 MeV, has a unique relaxation length for different pass-through materials: for example, lead, water, and air have relaxation lengths of 5, 50, and 10,000 cm respectively. The dose of gamma radiation received by unprotected human tissue can be significant because the dose is not greatly impacted by air molecules. The properties of the more common radioactive emissions are summarized in Table 15–2.

Table 15–2.   Properties of Ionizing Radiation

Particle or Wave	Mass in amu	Charge
alpha ($_2\text{He}^4$)	4	+2
beta (electron)	5.5×10^{-4}	−1
gamma (X-ray)	approx. 0	0
neutron	1	0
positron	5.5×10^{-4}	+1

When ionizing radiation is emitted from a nucleus, the nature of that nucleus changes: Another element is formed and there is a change in mass as well. This process may be written as a nuclear reaction. In such a reaction, both mass and charge must balance for reactants and products. For example, the beta decay of C-14 may be written:

$$_6\text{C}^{14} = {_{-1}}\text{e}^0 + {_7}\text{N}^{14}$$

The mass balance for this nuclear equation is

$$14 = 0 + 14$$

The charge balance is

$$6 = -1 + 7$$

A typical reaction for alpha decay is the first step in the Uranium-238 decay chain:

$$_{92}U^{238} = {}_2He^4 + {}_{90}Th^{234}$$

Thus when a nuclide emits a beta, the mass number remains unchanged and the atomic number increases by one. Beta decay is thought to be the decay of a neutron in the nucleus to a proton and an electron (beta particle), with emission of the beta from the decaying nucleus. When a nuclide emits an alpha, the atomic mass decreases by 4 and the atomic number decreases by 2. Gamma emission does not yield a change in atomic mass or atomic number.

Nuclear reactions can also be written for bombardment of nuclei with subatomic particles. For example, tritium (H^3) is produced by bombarding a lithium target with neutrons:

$$_0n^1 + {}_3Li^6 = {}_1H^3 + {}_2He^4$$

The reactions written above tell us nothing about the energy with which ionizing radiation is emitted. However, the emitted particle always has a certain amount of kinetic energy. This energy is lost in collisions with target atoms as the alpha, beta, neutron, and gamma radiations pass through different materials. If the material is human tissue, the energy gained (by the tissue) causes disordering in the chemical or biological structure of the tissue and may produce cell deaths or subsequent dysfunction of the cells. These effects are discussed later in this chapter.

Units for Measuring Radiation

Standard units have evolved that are used to measure radiation and its impact on material. As we see above, the damage to human tissue is directly related to the amount of energy deposited in the tissue by the alpha, beta, and gamma particles. This energy, in the form of ionization and excitation of molecules, results in heat damage to the tissue or even radiation burn. For this reason, many of the units used to measure radiation are related to energy units.

A *curie* (Ci) is a measure of total radioactivity or source strength and is equal to 3.7×10^{10} disintegrations/second—the radioactivity of one gram of the element Radium. The decay rate is measured in curies. The source strength in curies is not sufficient for a complete characterization of a source; the nature of the element (for example, Pu^{139}, U^{138}, Sr^{90}) and the type of emission (for example, alpha) are also necessary. Since 1986 the Bequerel (Bq) has also been used as a measure of source strength, 1 Bq = 1 disintegration per second; one curie = 3.7×10^{10} Bq.

The mass of material that disintegrates 3.7×10^{10} times per second varies widely from material to material. The relationship between the curies and the mass of a given radionuclide is given by:

$$Q = \frac{K_T M N^0}{W(3.7 \times 10^{10})}$$

where Q = number of curies

K_T = disintegration constant = $0.693/t_{1/2}$; that is, the fraction of atoms that decay per second

$t_{1/2}$ = halflife of the radionuclide, in seconds

M = mass of the radionuclide, in grams

N^0 = Avogadro's number, 6.03×10^{23} atoms/gram atomic weight

W = atomic weight of the radionuclide in grams/gram atomic weight

A complex relationship exists between the quantity of radiation, the curie, and the radiation dose rate, rad/sec. This relationship depends on the energy of the radiation, the type of radiation, the path of the radiation, and the amount of absorbing material between the emitter and the receptor.

A *rad* (for *radiation absorbed dose*) is that quantity of ionizing radiation that leads to absorption of 100 ergs/gram of absorbing material. A unit of dose whose use is increasing is the *Gray* (Gy). 1 Gy = 1 Joule/kg = 100 rad.

The *Roentgen* was for many years a standard unit of exposure corresponding to a quantity of gamma rays that deposit 87.7 ergs per gram of air at standard temperature and pressure. This unit is equivalent to the ejection of 1 electrostatic unit (esu) of charge in 1 cm^3 of air and describes the electromagnetic field associated with radioactive decay. The unit does not indicate where the gamma rays deposit their energy, however. For the environmental engineer interested in health effects of radiation, the locations of their energy deposit is also of concern.

The *dose equivalent,* measured in *rem* (for "Roentgen equivalent man"), addresses the following issue: all types of ionizing radiation do not produce identical biological effects for a given amount of energy delivered to human tissue. Radiation yielding higher specific ionization along its track will generally produce a greater effect, but the quantitative difference will depend on the tissue or organ and biological change selected for study. The dose equivalent is the product of the dose in rad and a *quality factor* (sometimes called *relative biological effectiveness,* or *RBE*), which describes the biological effect being considered.

Different types of ionizing radiation, even for a given amount of energy, produce different effects on living tissue. The quality factor takes into account the differing biological effects of alpha, beta, and gamma radiation. The standard for comparison is gamma radiation having a *linear energy transfer (LET)* in water of 3.5 keV/um, and a rate of 10 rad/min. Together with this standard, the quality factor can be defined as:

$$QF = rem/rad$$

Dose equivalent is also expressed in *sievert* (Sv).

$$1 \ Sv = 100 \ rem$$

and

$$QF = \frac{Sv}{Gy}$$

Table 15–3 gives sample quality factors for internal dose: the dose from radionuclides incorporated in human tissue.

Quality factors are determined for chronic, low-level doses of radiation, using effects that occur in an invidual during a lifetime of exposure. Acute, high-level doses produce different effects and have different quality factors.

Doses are also often expressed in terms of *population dose,* which is measured in *person-rem.* The population dose is the product of the number

Table 15–3.   Sample Quality Factors for Internal Radiation*

Internal Radiation	Quality Factor
α	10.0
β and γ	
$E_{max} > 0.03$ Mev	1.0
$E_{max} < 0.03$ Mev	1.7

*E_{max} refers to the maximum level of emissions by the β and γ source.

Table 15–4.   Estimates of Annual Whole-Body Dose Rates in the United States

Source	Average Whole-Body Dose mrem/yr	Annual Population Dose 10^6person-rems
Natural sources		
Cosmic radiation	44	
Internal (K^{40})	18	
External alpha		
(minerals)	30	
External beta	10	
Subtotal	102	21
Anthropogenic sources		
Diagnostic X-ray	72	15
Radiopharmaceuticals	1	0.2
Global fallout	9	2
Occupational exposure	1	0.2
Nuclear power	0.03	0.01
Miscellaneous	2	0.5
Subtotal	85	18
TOTAL	187	39

of people affected and the average dose in rem. That is, if a population of 100,000 individuals receives an average whole-body dose of 0.5 rem, the population dose is 0.5×10^5 person-rem or 50,000 person-rem. The utility of the population dose concept will become evident in the section dealing with health effects.

Table 15–4 gives some average radiation doses in the United States. It should be remembered that these are rather imprecise estimates.

Table 15–4 does not include a background source of ionizing radiation whose importance was only recently adequately assessed: Radon gas (Rn^{222}). Rn^{222} is a daughter of U^{238}, which is ubiquitous in soil, rock, building materials, and so forth. Since Rn^{222} is chemically an inert gas, it is not trapped chemically. It is present not only in uranium ore but also in ordinary soil and in a number of rock formations in the United States, most notably in the Piedmont plateau and other parts of Appalachia. Rn^{222} from soil can accumulate in the basements of buildings and in occupied areas of buildings if ventilation is not adequate. Rn^{222} itself has a short halflife (3.4 days) and, being inert, is not absorbed or trapped in the lung. During its transit through the human lung, however, it can decay to Po^{218} and eventually to Po^{214}. These nuclides are adsorbed to the lung surface. The National Academy of Sciences now estimates that the annual average dose from inhaled alpha emitters is 2500 mrem per year, and that this dose is almost entirely due to the radio progeny Po^{218} and Po^{214}. Rn^{222} is probably the single greatest "background" source of human exposure.

Measuring Radiation

Devices have been developed to measure the radiation dose, dose rate, or the quantity of active material that is present. The particle counter, the ionization chamber, photographic film, and the thermoluminescent detector are four methods widely used in the field.

Counters are designed to note the movement of single particles through a defined volume. Gas-filled counters collect the ionization produced by the radiation as it passes through the gas and amplify it to produce an audible pulse. Counters are typically used to determine the radioactivity present by measuring the number of particles of protons that are emitted.

Ionization chambers basically consist of a pair of charged electrodes that collect ions formed within their respective electrical fields. Chambers are generally designed to determine dose or dose-rate measurements because they provide an indirect representation of the energy deposited in the chamber.

Photographic film darkens if exposed to radiation and is a useful indicator of the presence of radioactivity. Such film is often used for determining personnel exposure and making other dose measurements where a long record of dose is necessary or a permanent record of dose is required.

Thermoluminescent detectors (TLD) are crystals, such as NaI that can be excited to high electronic energy levels by ionizing radiation. The energy is then released as a short burst or flash of light that can be detected by a photocell or photomultiplier. TLD systems are replacing photographic film in personnel dosimeters because they are more sensitive and consistent. Liquid scintillators (organic phosphors) are used in biomedical applications.

HEALTH EFFECTS

When alpha and beta particles and gamma radiation penetrate living cells or any other matter, they transfer their energy through a series of collisions with the atoms or nuclei of the receiving material. Many molecules

Table 15–5.   Representative LET Values

Radiation	Kinetic Energy (MeV)	Av. LET (kEV/u)	Dose Equiv.
X-rays	0.01–0.2	3.0	1.00
Gamma rays	1.25	0.3	0.7
Electrons (beta)	0.1	0.42	1.0
	1.0	0.25	1.4
Alpha particles	0.1	260	—
	5.0	95	10
Neutrons	thermal		4–5
	1.0	20	2–10
Protons	2.0	16	2
	5.0	8	2

are damaged in the process as chemical bonds are broken and electrons are lost (ionization). In fact, energy is lost all along the path of the radiation, and a measure of this rate of *linear energy transfer* (LET) is the density of ionization activity along this path. The more ionization that is observed, the higher the intensity of biological damage to the cells. Table 15–5 gives some typical LET values. Since LET is a measure of the energy lost by the ionizing radiation with each successive collision, charged particles may have a higher LET than uncharged particles of the same size and may penetrate further. Values given for the dose equivalent vary widely with different measures of biological damage.

Biological effects from these penetrations can be grouped as *somatic* and *genetic*. Somatic effects are the impacts on individuals directly exposed to the radiation and include damage to the circulatory system, carcinogenesis, and decrease in organ function because of cell killing. Genetic effects occur because ionizing radiation damages the genetic material of the cell and can cause chromosome breakage. Genetic effects are not evident in the individual receiving the radiation but are transferred to that individual's offspring and descendants.

Radiation sickness (circulatory system breakdown, nausea, hair loss) and resulting death are acute somatic effects occurring after very high exposure as from a nuclear bomb or intense radiation therapy. The accident at the Chernobyl nuclear generating plant in the USSR in 1986 resulted in about 30 deaths from acute effects during the 2 months after the event. These individuals received estimated gamma-radiation whole-body doses of approximately 400 rads. Most mammals appear to have the same degree of sensitivity to the LD_{50} (lethal dose resulting in a 50 percent kill within 30 days). Lower doses, but above 100 rads, will lead to vomiting, diarrhea, and nausea in humans; hair loss is generally observed within two weeks after an individual has been exposed to 300 rads or more.

Ionizing radiation presents a good example of the different effects of acute high doses and chronic low doses. Construction of dose-response curves depends on retrospective studies of atom bomb survivors, industrial and occupational exposures, and a few studies of very high medical exposures. Since low doses produce effects (cancer, genetic effects) that have very long latency periods, good data does not exist for low doses, and most dose-response curves have been extrapolated from high-dose data. The population exposed in the Chernobyl accident is being monitored closely, but no chronic effects have been observed as yet.

There is no threshold for cell damage from ionizing radiation. Thus for a long time, low-dose responses were linear extrapolations of high-dose data. Recent thorough reexamination of all data on radiation effects indicates that the dose response curve is probably linear at low doses and quadratic at high doses.

Unless there is a nuclear war, only low-LET radiation need be considered as an environmental pollutant. Since such radiation is a carcinogen, there is assumed to be no threshold for damage, and risk analysis is necessary. As effects of exposure have been expressed over the years since the first atomic bomb, risk estimates have become more refined. Tables 15–6 and 15–7 give a variety of risk estimates for effects from exposure to low-LET radiation including the estimates presently used by the U.S. Environmental Protection Agency (EPA). The uncertainty in these numbers is estimated to be ±250 percent. As we have seen in a previous chapter, the number of cancer fatalities in the United States per year per 10^6 persons is 1930. In other words, exposure to an additional rad per year would increase the number of fatal cancers per million population by about 15 percent. The uncertainty in these numbers is estimated to be ±300 percent. By comparison, the current incidence of genetic defects in the United States per 10^6 liveborn children is 107,000.

Table 15–6.   Fatal Somatic Risk from Lifetime Exposure to
1 Rad/Year Low LET Radiation

Year	Dose Model	Effects per 10^6 Person-Rad
1972		667
1980	Linear	403
1980	Quadratic	169
1977	Quadratic	75–175
1985	Linear	280

Table 15–7.   Genetic Risk per 10^6 Liveborn from Exposure
to 1 Rem/Generation Low LET Radiation

Year	First Generation	All Generations
1956		500
1972	12–200	60–1500
1977	63	300
1980	89	320
1982	22	149
1985	20–90	260–1110

There is a documented risk from low levels of low LET ionizing radiation, but it is apparently small and very uncertain. The health effects of a given dose of radiation depend on a large number of factors, including

- magnitude of the absorbed dose
- type of radiation
- penetrating power of the radiation
- sensitivity of the receiving cells and organs
- rate at which the dose is delivered
- proportion of the cell/organ/human body exposed

To protect public health, one should act to minimize unnecessary radioactive exposure.

SOURCES OF RADIOACTIVE WASTE

There are a number of sources of radioactive waste: the nuclear fuel cycle, radiopharmaceutical manufacture and use, biomedical research and applications, and a number of industrial uses. The behavior of radionuclides is determined by their physical and chemical properties; radionuclides may exist as gases, liquids, or solids and may be soluble or insoluble in water or other solvents.

Until 1980, there was no classification of radioactive wastes. In that year, the U.S. Nuclear Regulatory Commission (NRC) classified radioactive wastes into the following categories:

High-level waste (HLW).   HLW includes two categories: spent nuclear fuel from nuclear reactors and the solid and liquid waste from reprocessing spent fuel for defense purposes. In the United States, only military irradiated fuel is reprocessed, to recover plutonium and fissile uranium.

The NRC reserves the right to classify additional materials as HLW as necessary.

Uranium mining and mill tailings.   The pulverized rock and leachate from uranium mining and milling operations.

Transuranic (TRU) waste.   Radioactive waste that is not HLW but that contains more than 100 nanocuries per gram of elements heavier than uranium (the elements with atomic number higher than 92). Most TRU waste in the United States is the product of defense reprocessing.

By-product material.   Any radioactive material, except fissile nuclides, that is produced as waste during plutonium production or fabrication.

Low-level waste (LLW).   LLW includes everything that is not included in one of the other four categories. In order to assure appropriate disposal, the NRC has designated several classes of LLW:

> *Class A* contains only short-lived radionuclides or extremely low concentrations of longer-lived radionuclides and must be chemically stable. Class A waste may be disposed of in landfills without particular stabilization as long as it is not mixed with other hazardous or flammable waste. Most trash is Class A waste.

> *Class B* contains higher levels of radioactivity, must be physically stabilized before transportation or disposal, and cannot contain free liquid.

> *Class C* is waste that will not decay to acceptable levels in 100 years and must be isolated from the environment for 300 years or more. Power plant LLW is in this category.

LLW is not necessarily less radioactive than HLW, and may even have a higher specific activity (Curies/gm). The distinguishing feature of LLW is that it contains virtually no alpha emitters.

The Nuclear Fuel Cycle

The nuclear fuel cycle generates radioactive waste at every stage. Mining and milling generate the same sort of waste that any mining and milling operations generate except that this waste is radioactive. Mining and milling dust must be stabilized to keep it from dispersing, and leachate must be prevented from contaminating waterways and groundwater.

Partially refined uranium ore (called "yellowcake" because of its bright yellow color) is then enriched in the fissile isotope U^{235}, and nuclear fuel is fabricated. Mined uranium is more than 99 percent U^{238}, which is not fissile, and approximately 0.7 percent U^{235}. The concentration of U^{235} is increased to about 3 percent by converting to UF_6. The lighter isotope is then separated by gaseous diffusion. UF_6 enriched in U^{235} is then converted to UO_3 and fabricated into fuel. Enrichment and fabrication produce TRU waste.

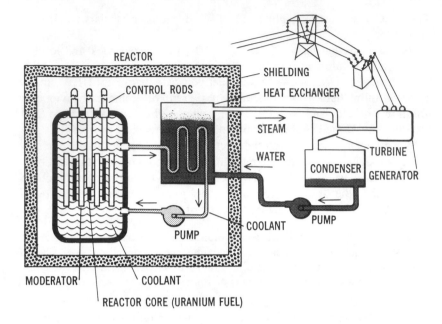

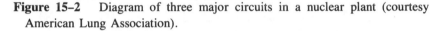

Figure 15-2   Diagram of three major circuits in a nuclear plant (courtesy American Lung Association).

The nuclear fuel is then inserted into a nuclear reactor core where a controlled fission reaction produces heat, which in turn produces pressurized steam for electric power generation. The steam system, turbines, and generators in a nuclear power plant are essentially the same as those in any thermal electric power plant (like a coal plant). The difference between nuclear and fossil fuel in the generation of electricity is in the evolution of the heat that drives the plant.

Figure 15-2 is a diagram of a typical pressurized-water nuclear reactor. Commercial reactors in the United States are either pressurized water reactors in which the water that removes heat from the nuclear reactor core (the "primary coolant") is under pressure and does not boil, or boiling water reactors in which the primary coolant is permitted to boil.[3] In any case, the primary coolant transfers heat to the steam system (the "secondary coolant") by a heat exchanger that assures complete physical

---

[3]One exception is the Fort St. Vrain reactor in Colorado, in which the primary coolant is helium gas.

isolation of the primary from the secondary coolant. A third cooling system provides water from external sources to condense the spent steam in the steam system.

All thermal electric power generation produces large quantities of waste heat. Fossil fuel electric generating plants are at best about 42 percent thermally efficient: That is, 42 percent of the heat liberated by combustion in the boiler is converted to electricity, and 58 percent is simply dissipated into the environment. By comparison, nuclear plants are 32 percent efficient at best.

Reprocessing Waste

In the United States, plutonium for weapons is produced by irradiating U^{238} with neutrons (in the reaction given above) in military breeder reactors. The plutonium, along with uranium and neptunium, is then extracted by dissolving the entire irradiated fuel element in nitric acid and then extracting with tributyl phosphate. Further partition and selective precipitation result in recovery of plutonium, uranium, neptunium, strontium, and cesium. The fissile isotopes of plutonium and uranium are categorized as *special nuclear material;* the remainder of these are considered by-product material.

Both the acid solvent, which is ultimately neutralized, and the organic extraction solvents contain high concentrations of radioisotopes and are classified as HLW. This process also yields TRU and low-level wastes.

The only use for plutonium in the United States is in weapons manufacture; commercial spent fuel is not reprocessed. In France, however, plutonium is produced (also by neutron reaction with U^{238}) in the Superphenix breeder reactor for use in nuclear electric power generation.

Other Reactor Waste

The primary and secondary coolants in a nuclear generating plant pick up considerable radioactive contamination through controlled leaks. Contaminants are removed from the cooling water by ion-exchange columns. The loaded columns are Class C low-level radioactive waste. Class A and B wastes are also produced in routine cleanup activities in nuclear reactors.

Eventually, the reactor core and the structures immediately surrounding it become very radioactive, primarily by neutron activation, and the reactor must be shut down and decommissioned. Although the two small reactors now being decommissioned are being treated like LLW, the ultimate implications of decommissioning are not yet clear.

Other Sources of Radioactive Waste

The increasingly widespread use of radioisotopes in research, medicine, and industry has created a lengthy list of potential sources of other radioactive waste. Sources range from a large number of laboratories using small quantities (a few isotopes) to large medical and research laboratories where many different isotopes are produced, used, and wasted in large volumes.

Liquid scintillation counting has become an important biomedical tool and produces large volumes of waste organic solvents such as toluene, which have a low specific activity. The long-lived contaminants of liquid scintillators are tritium and C^{14} for the most part. These wastes are characteristic of mixed wastes: mixtures of hazardous and radioactive waste. Their chemical nature as well as their radioactivity poses disposal problems.

Naturally occurring radionuclides and those inadvertently released in times past can also pose threats to public health. An outstanding example is the buildup of Rn^{222} in homes and commercial buildings with restricted air circulation. Rn^{222} is a member of the uranium decay series, and is thus found ubiquitously in rock. Chemically, Rn^{222} is an inert gas like helium. When uranium-bearing minerals are crushed or machined, Rn^{222} is released; there is even a steady release of Rn^{222} from rock outcrops. Buildings that are insulated to prevent convective heat loss often have too little air circulation to keep the interior purged of Rn^{222}. Although Rn^{222} has a short halflife, it decays to much longer-lived metallic radionuclides.

Coal combustion, copper mining, and phosphate mining release isotopes of uranium and thorium into the environment. K^{40}, C^{14} and H^3 are found in many foodstuffs. Although atmospheric nuclear testing was discontinued many years ago, fallout from past tests continues to enter the terrestrial environment.

A radionuclide is an environmental pollutant because of a combination of properties:

- Halflife
- Chemical nature and properties: A radioactive isotope will behave chemically and biochemically exactly like stable (nonradioactive) isotopes of the same element.
- Abundance

From an *air transport* standpoint, all nuclear waste handling and disposal facilities fall into two general categories: (1) those that have a planned and predictable discharge to the atmosphere and (2) those where any discharge would be purely accidental. Boiling-water reactors, fuel processing plants, and a large number of atomic experimental systems—particularly military—fit into the first category and on occasion also into the second category. Good examples of the second category are well contained pressurized water reactors and long-term storage facilities.

The atmospheric pathways from both types of discharges have received a great deal of attention because of the analogies between radioactive emission, sulfur dioxides, and inert gases generated by automobiles and smoke stacks across the nation. Airborne radioactivity is available for inhalation and/or disposition on water supplies or food chain land and thus impacts public health. Chapters 17 and 18 discuss air pollution and meteorology, and much of that discussion is applicable to the study of radioactive particles.

Water transport occurs whenever radionuclides in surface or subsurface soil erode or leach into a watercourse or whenever fallout occurs from the atmosphere. The radioactive waste is transferred from its solid or liquid state into groundwater or surface water supplies where the radionuclides can enter the human food chain or drinking water. Little is known, however, about the leachability of radionuclides, and only gross estimates can be made about surface water transport and disposition. As with any such radioactive material transport equations, the halflife of the isotopes must be considered. Chapters 4 and 9 address water transport considerations that are somewhat applicable to the problems associated with radioactive waste.

Radioactive isotopes also move *from the land* into the human food chain. Plant uptake mechanisms have virtually no way of screening an element that is radioactive; thus plants generally collect a deposit of radionuclides in their structure if the radionuclides are present with the water, phosphorus, nitrogen, and trace elements necessary for plant growth. Rain and wind also act to deposit the hazardous material on the roots and leaves of the plants, all to be passed along to whatever or whomever eats the plants. Cows eating contaminated corn, for example, can pass radioactivity along to milk drinkers.

Transuranic wastes last for long periods of time within the environment, but for the most part are strongly held by soil particles. They are not easily translocated through most food chains although some concentration does pass through certain aquatic food systems. In addition, they pose only a slight biological hazard to humans because adsorption in the gastrointestinal track is limited. The greater potential hazard from TRU wastes is from inhalation of dust particles containing the materials since a large fraction is generally retained in the human lung.

In summary, the variety of chemical characteristics displayed by radioactive materials allow them to be transported through the environment by a number of different pathways, making the management of such wastes especially troublesome.

RADIOACTIVE WASTE MANAGEMENT

The objective of the environmental engineer is to prevent the introduction of radioactive materials into the biosphere during the effective lifetime (about 20 halflives) of these materials. Control of the potential direct impact on the human environment is necessary but not sufficient because radionuclides can be transmitted through water, air, and land pathways for many years and in some cases for many generations. Some radionuclides that are currently treated as waste may be retrieved and recycled by future reprocessing, but such recycling creates its own waste stream. Much radioactive waste is not amenable to recycling. It can be treated only by isolating it from pathways to the human food chain and environment until its radioactivity no longer poses a threat. Isolation requirements differ for the different classes of radioactive waste. With the given halflives of many radioisotopes, it is difficult to imagine any technology that truly offers ultimate disposal for these wastes; thus we will think in terms of long-term storage: 10, 100, or 10,000 years or longer. Many of the issues discussed in Chapter 14 are applicable to the radioactive waste problem as well.

High-Level Radioactive Waste

The following options have been considered for long-term disposal of HLW:

1. land disposal
    - burial in very deep holes
    - burial on an inaccessible island
    - deep (mined) geologic disposal
    - liquid injection into geologic formations
    - rock melting
2. subseabed disposal
3. disposal in polar ice sheets
4. disposal in space
5. transmutation into shorter-lived or stable nuclides

Only two of these options appeared to warrant further investigation: mined geologic and subseabed disposal. After preliminary investigation, the subseabed option has been discontinued because the waste could not be retrieved if this ever became necessary or prudent. Mined geologic disposal appears to be the only available option.

In order to store radioactive waste with a reasonable degree of assurance that it will not be dispersed into the environment, a three-stage barrier has been proposed. The first barrier would be provided by the waste form itself: The optimum waste form would be radioactive material dispersed in a glass matrix or vitrified. The second barrier would be provided by an engineered system, which includes the waste packaging. The third barrier would be the geological (rock) matrix.

Vitrification is planned for defense HLW. Since commercial spent fuel is not reprocessed, however, it will be stored in a geological repository in the form in which it leaves the reactor, as spent fuel rods, which are combined in heavy stainless steel casks in the engineered barrier system.

When the fissile uranium in a fuel rod, in a reactor is about 75 percent used up, the rods are ejected into a very large pool of water where they remain until the short-lived nuclides have decayed completely and the rods are thermally cooler. Initially, this phase was intended to be about six months long, but the spent fuel has remained in storage pools as long as ten years in some cases because there is no repository for it. After sufficient decay and cooling, the spent fuel will be loaded into casks and emplaced in a repository.

Investigations for a geological repository began in 1972. Salt is the most thoroughly investigated substance because:

- Salt mining technology is well developed and storage sites can be constructed.
- Salt deposits tend to have a high plasticity and thus have a tendency to seal themselves if fractures are created by major movements in the earth's crust.
- Salt deposits have low permeability and are essentially sealed from groundwater and surface water supplies.
- Salt has a high thermal conductivity, which helps dissipate heat that builds in waste containers.
- Salt formations have a high structural strength with the ability to withstand effects of heat and radiation.

Salt has some disadvantages, however. No salt deposit is free of brine inclusions, and these tend to migrate toward heat sources, which the emplaced radioactive waste would provide. Salt also contains some fossil water.

The Nuclear Waste Policy Act of 1982 mandated two geological repositories and further mandated geological diversity since the behavior of any single medium over such long time periods can only be guessed at. This law was amended in 1987 to postpone investigation of a second repository indefinitely. Three rock types—salt, volcanic tuff, and basalt—were being investigated for suitability for the first repository but only volcanic tuff is presently under study. The repository is scheduled to begin accepting waste in 2012 and to close, sealing the waste permanently and irretrievably, in 3012.

Except for salt, the media investigated are all hard rock, which cracks and fissures under thermal and chemical stress. Radioactive emissions may even assist in the cracking. Repository integrity thus depends on keeping radionuclides out of ground water aquifers if any are present; this poses potential problems for the hard rock repositories.

TRU Waste

Since 1979, a salt formation near Carlsbad, New Mexico, has been under investigation as a radioactive waste repository. The Waste Isolation Pilot Project (WIPP) was found to be unable to withstand the thermal stress from spent fuel and will serve instead as the TRU waste repository for the United States. WIPP is scheduled to begin receiving waste in the early 1990s.

CONCLUSION

There are now about 100 operating nuclear electric generating plants in the United States, and no more are being ordered or designed. Plutonium production reactors are nearing the end of their useful lives, and no more are planned at present. Even if the entire nuclear industry were to be shut down, however, the waste would still be with us. We have just embarked on thorough investigations of disposal options for radioactive waste, and we can anticipate steps toward resolution in the next decade.

Should the nuclear industry shut down? Should the currently operating plants finish their useful lives of 30–40 years and not be replaced? In this context, we must realize that 20 percent of the electricity in the United States today is nuclear-generated. In some states, the share is as high as 30 percent (Massachusetts) and in one instance better than 60 percent (Illinois). It is unlikely that we will be able to replace this generating capacity with solar, wind, and geothermally generated electricity in the next 30 years if indeed we can do it at all. Energy conservation can perhaps increase supplies by 5 percent.

The other future option for electric generation is coal combustion, which has far reaching environmental effects. We must make a thorough and objective comparison of the environmental and public health costs and benefits of these options before making any decisions that apply unilaterally to nuclear power. It is wise to remember that generating electricity, no matter how it is done, causes permanent, irrevocable environmental damage on a scale directly proportional to the power produced.

Military plutonium production is slated to be curtailed by the middle of the 1990s. Open debate is needed on the future of the nuclear arsenal.

If nuclear weapons and nuclear power go away, will nuclear medicine and radiopharmaceuticals go away as well? We should be very wary of making hasty, "throwing-the-baby-out-with-the-bath-water" decisions.

In the past 40 years, the pendulum of public opinion has swung from the pronuclear to the antinuclear extreme but seems to be reversing. Politics follows public opinion, but political decisions about nuclear power and nuclear weapons are not always wise ones. The role of environmental scientists, now more than ever, is to inform political decision makers.

PROBLEMS

15.1 Qualitatively rank the four types of long-term storage (land, ocean, ice, and spaceship) according to each of the following criteria:

- environmental protection (water, air, and surface land)
- direct protection of public health
- retrievability of waste
- time available for undisturbed storage
- storage capacity
- cost of disposal

Discuss your rankings in detail and make general recommendations for the Governor's Task Force on radioactive waste disposal in your home state.

15.2 Discuss the *environmental* impacts of a nuclear power plant, including water, air, and land quality considerations. Include discussions of thermal pollution and impacts on the aquatic environment.

15.3 Assume your sister-in-law, 7 months pregnant, has a toothache. Outline the benefits and costs associated with these three courses of action:

- no dentist, no treatment
- dentist visit, treatment aided with X-rays
- dentist visit, treated without the aid of X-rays

What would you advise?

15.4 Show that after ten halflives about 0.1 percent of the initial radioactive material is left and that after twenty halflives 10^{-4} percent is left.

15.5 Fe^{55} has a halflife of 2.4 years. Calculate the disintegration constant. If there are 16,000 curies of FE^{55} in a reactor core vessel, how much will be left after 100 years?

15.6 I^{131} is used in diagnosis and treatment of thyroid diseases. How should a hospital dispose of waste I^{131}? Should this waste be disposed of in a shallow land burial site? Are other disposal methods available and adequate?

15.7 Write nuclear reactions for the following, identifying the product elements and particles:

  a. Fusion of two deuterium atoms to form He^3.
  b. beta decay of Sr^{90}
  c. beta decay of Kr^{85}
  d. neutron emission from Kr^{87}
  e. decay of Th^{230} to Ra^{226}

15.8 How many grams of Cs^{137} are produced in a 750 MW_t power plant? How many grams would be left if this amount were allowed to decay for 100 years?

15.9 From Table 15–4, calculate the average energy absorbed per year by an individual in the United States. Assume that all beta and gamma radiation is low energy (0.03 MeV).

15.10 The mixture of uranium and its daughter elements that is found in old uranium deposits is called *secular equilibrium:* the equilibrium amounts of U^{238} and each daughter element. If you begin with pure U^{238}, how long does it take until secular equilibrium is reached?

15.11 The environmental impact statement for radioactive waste transportation limits the dose a truck driver can receive to 4.3 mrem. For how many hours would a truck driver receive this dose until his or her total dose would equal the annual dose from natural radiation?

15.12 The EPA limits the allowed dose from airborne radioactivity to 25 mrem/year. Estimate the number of somatic effects that might occur in a population of 50,000 who receive this dose over a 70-year period in addition to the background dose. Make your assumptions explicit.

15.13 What do you think should be the upper limit for annual occupational radiation dose? Why?

LIST OF SYMBOLS

α = alpha radiation

β = beta radiation

γ = gamma radiation

amu = atomic mass unit

esu = electrostatic unit

K_b = fraction of atoms that disintegrate per unit time

K_T = fraction of atoms that disintegrate per second = $0.693/_{\frac{1}{2}}L$

$t_{\frac{1}{2}}$ = halflife of a radionuclide, in seconds

LD_{50} = lethal dose 50

LDC_{50} = lethal dose concentration 50

LET = linear energy transfer

M = mass of radionuclide, grams

MeV = 10^6 electron volts

N_0 = number of radioactive atoms within a material

N^0 = Avogadro's number, 6.02×10^{23} atoms/gram atomic weight

Q = number of Curies

rem = Roentgen equivalent man

t = time

TRU = transuranic (long-lived isotopes)

W = atomic weight of a radionuclide, grams/gram atomic weight

Appendix A:
Radioactive Decay

An atom that is radioactive has an unstable nucleus. The nucleus moves to a more stable condition by emitting an alpha or beta particle; this emission is frequently accompanied by emission of additional energy in the form of gamma radiation. This emission is called *radioactive decay*. Radioactive decay follows the laws of chance, and the rate of decay, or rate of decrease of the number of radioactive nuclei, can be expressed as an ordinary first-order reaction:

$$dn = -K_b dt$$

where   N   = the number of radioactive nuclei
K_b = a proportionality factor called the disintegration constant; K_b has the units 1/time

Integrating this equation, letting $N = N_0$ at time $= 0$, we get the classical exponential radioactive decay equation:

$$\int_{N_0}^{N} \frac{dN}{N} = -\int_{0}^{t} K_b dt$$

$$\ln\frac{N}{N_0} = -K_b t$$

Rearranging gives us:

$$\frac{N}{N_0} = \exp(-K_b t)$$

The data points in Figure 15A–1 correspond to this equation.

After a specific time period, $t = \frac{1}{2}L$, the value equals one-half the previous N. That is, at the end of some time period defined as $t_{\frac{1}{2}}$, half the radioactive atoms have disintegrated. This halflife is determined from the above equation:

$$t_{\frac{1}{2}} = \frac{\ln 2}{K_b} = \frac{0.693}{K_b}$$

Looking at Figure 15A–1, we see that N does not become zero at $t = 2 \cdot t_{\frac{1}{2}}$. In fact N becomes $\frac{1}{4}N_0$. We therefore see that the equation

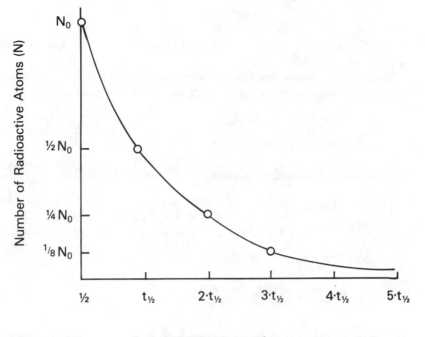

Key:

$t_{1/2}$ = one half-life

$2 \cdot t_{1/2}$ = two half-lives

Figure 15–A1   General description of radioactive decay.

$$\frac{N}{N_0} = e^{-K_b t}$$

is so constructed that N never becomes zero for any finite time period; for every halflife that passes, the number of atoms is halved.

Example 15A.1

An engineer, with the assistance of a chemist, prepares 10.0 g of pure $_6C^{11}$. This notation represents carbon with 6 protons and a total mass of 11 atomic mass units (amu, the number of protons plus the number of neutrons). (Conversion: 1 amu = 1.66×10^{-24} g.) The radioactive decay of this atom is

$$_6C^{11} \rightarrow {}_1e^0 + {}_5B^{11}$$

If the halflife is 21 minutes, how many grams of carbon are left 24 hours after the preparation?

The equations above describing the halflife refer to number of atoms so we must initially calculate the number of atoms in 1 g of $_6C^{11}$

$$\frac{1 \text{ atom } _6C^{11}}{11.0 \text{ amu}} \times \frac{1 \text{ amu}}{1.66 \times 10^{-24} \text{ g } _6C^{11}} = \frac{1 \text{ atom } _6C^{11}}{18.3 \times 10^{-24} \text{ g } _6C^{11}}$$

Now compute the number of atoms of $_6C^{11}$ in 10 g

$$\frac{1 \text{ atom } _6C^{11}}{18.3 \times 10^{-24} \text{ g } _6C^{11}} \times 10 \text{ g } _6C^{11} = 55 \times 10^{22} \text{ atoms } _6C^{11}$$

Now apply the equation for $_{\frac{1}{2}}L$:

$$K_b = \frac{0.693}{t_{\frac{1}{2}}} = \frac{0.693}{21 \text{ min}} = 33 \times 10^{-3} \text{ min}^{-1}$$

Following convention to avoid dealing with negative logarithms, we can rearrange our halflife equation as:

$$-K_b t = \ln \frac{N}{N_0} = -\ln \frac{N_0}{N}$$

or

$$K_b t = \ln \frac{N_0}{N}$$

Now realizing that we must express t and K_b so that $(K_b t)$ is unitless, we must include the number of minutes in a day in our calculations:

$$K_b t = (33 \times 10^{-3} \text{ min}^{-1})(1440 \text{ min}) = 47.5$$

and

$$\ln \frac{N_0}{N} = 2.303 \log \frac{N_0}{N}$$

So we have

$$2.303 \log \frac{N_0}{N} = 47.5$$

or

$$\log \frac{N_0}{N} = 20.6$$

Applying the properties of logarithms:

$$\frac{N_0}{N} = (\text{antilog } 0.6) \times (\text{antilog } 20)$$

where we get:

$$\frac{N_0}{N} = 4.0 \times 10^{20} \text{ or } N = \frac{N_0}{4.0 \times 10^{20}}$$

This gives us

$$N = \frac{55 \times 10^{22}}{4.0 \times 10^{20}} = 14 \times 10^2 \text{ atoms } {}_6C^{11}$$

After 1 day there are approximately 1400 atoms of ${}_6C^{11}$ remaining. To convert this to grams:

$$(14 \times 10^2 \text{ atoms } {}_6C^{11}) \times \left(\frac{18.3 \times 10^{-24} \text{ g}}{\text{atoms } {}_6C^{11}} \right) = 250 \times 10^{-22} \text{ g } {}_6C^{11}$$

Chapter 16

Solid and Hazardous Waste Law

Laws controlling environmental pollution are discussed in this text in terms of their evolution from the courtroom through Congressional committees to administrative agencies. The gaps in common law are filled by statutory laws adopted by Congress and state legislatures, and implemented by administrative agencies such as the U.S. Environmental Protection Agency (EPA) and state departments of natural resources. For several reasons, this evolutionary process was particularly rapid in the area of solid waste law.

For decades, and in fact centuries, solid waste was disposed of on land no one really cared about. Municipal refuse was historically trucked to a landfill in the middle of a woodland, and industrial waste—often hazardous—was generally "piled out back" on land owned by the industry itself. In both cases, environmental protection and public health were not perceived as issues. Solid wastes were definitely out of sight and neatly out of mind.

Only relatively recently has public interest in solid waste disposal sites reached the level of concern that equals that for water and air pollution. Some local courtrooms and zoning commissions have dealt with siting selected disposal facilities over the years, but most decisions simply resulted in the city or industry hauling the waste a little farther away from the complaining public. At these remote locations, solid waste was not visible like the smokestacks that emitted pollutants into the atmosphere and the pipes discharging wastewater into the rivers. Pollution of the land from solid waste disposal facilities was and is a much more subtle type of pollution.

The passage of federal and state environmental statutes reflects this naivety. Initially the highly visible problem of air pollution was addressed in a series of clean air acts. Then the next most visible pollution was confronted in the water pollution control laws. Finally, as researchers dug deeper into environmental and public health concerns, it was realized that even obscure landfills and holding lagoons had the potential to significantly pollute the land and impact public health. Subsurface and surface water supplies, and even local air quality, are threatened by solid waste.

This chapter addresses solid waste law in two major sections: nonhazardous and hazardous waste. Radioactive waste is discussed in Chapter 15. The division reflects the regulatory philosophy for dealing with these two very distinct problems. Realizing the general lack of common law in this area, we move directly into statutory controls of solid waste.

NONHAZARDOUS SOLID WASTE

The most significant solid waste disposal regulations were developed under the Resource Conservation and Recovery Act (RCRA) of 1976. This federal statute, which amended the elementary Solid Waste Disposal Act of 1965, reflected the concerns of the public in general and Congress in particular about (1) protecting public health and the environment from solid waste disposal, (2) filling the loopholes in existing surface water and air quality laws, (3) ensuring adequate land disposal of residues from air pollution technologies and sludge from wastewater treatment processes, and (4) promoting resource conservation and recovery.

The EPA implemented RCRA in a manner that reflected these concerns. Disposal sites were catalogued as either landfills, lagoons, or landspreading operations and the adverse effects of improper disposal were grouped into eight categories of impacts

- *Floodplains* were historically prime locations for industrial disposal facilities because many firms elected to locate along rivers for power generation and/or transportation of process inputs or production outputs. When the rivers flood, the disposed wastes are washed downstream, impacting water quality.
- *Endangered species* may be threatened directly by habitat destruction or kills as the disposal site is being developed and operated.
- *Surface water quality* can also be impacted by certain disposal practices. Without proper controls of runoff and leachate, rainwater has the potential to transport pollutants from the disposal site to nearby lakes and streams.
- *Groundwater* impacts are of great concern because about half the nation's population depends on groundwater for water supply.
- *Food chain crops* can be adversely affected by landspreading or solid waste, which can impact on public health and agricultural productivity.

Food chain crops often accumulate and concentrate trace chemicals and increase the dangers to public health.
- *Air quality* can be degraded by pollutants emitted from waste decomposition such as methane and can cause serious pollution problems downwind from a disposal site.
- *Health* and *safety* of onsite workers and the nearby public can also be directly impacted by explosions, fire, and bird hazards to aircraft.

The EPA guidelines spell out the operational and performance requirements to eliminate these eight types of impacts from solid waste disposal.

Under operational standards, technologies, designs, and/or operating methods are specified to a degree that theoretically ensures the protection of public health and the environment. Any or all of a long list of operational considerations could go into a plan to build and operate a disposal facility: type of waste to be handled, facility location, facility design, operating parameters, and monitoring and testing procedures. The advantage of operational standards is that the best practical technologies can be utilized for solid waste disposal, and the state agency mandated with environmental protection can easily determine compliance with a specified operating standard. The major drawback is that compliance is generally not measured by monitoring actual effects on the environment but rather as an either/or situation where the facility either does or does not meet the required operational requirement.

On the other hand, performance standards are developed to provide given levels of protection to land/air/water quality around the disposal site. Determining compliance with performance standards is not easy because the actual monitoring and testing of groundwater, surface water, and land and air quality are costly, complex undertakings.

HAZARDOUS WASTE

Common law is particularly not well developed in the hazardous waste area. The issues discussed in Chapter 14 are newly recognized by society, and little case law has had a chance to develop. Since the public health impacts of solid waste disposal in general and hazardous waste disposal in particular were so poorly understood, few plaintiffs ever bothered to take a defendant into the common law courtroom to seek payment for damages or injunctions against such activities. Hazardous waste law is therefore discussed here as statutory law, specifically in terms of the compensation for victims of improper hazardous waste disposal and efforts to regulate the generation, transport, and disposal of hazardous waste.

Historically, federal statutory law was generally lacking in describing how victims should be compensated for improper hazardous waste disposal. A complex, repetitious, confusing list of federal statutes was the only recourse for victims.

The federal Clean Water Act only covers wastes that are discharged to navigable waterways. Only surface water and ocean waters within 200 miles from the coast are considered, and a $5 million revolving fund is set up by the act and administered by the Coast Guard. Fines and charges are deposited into the fund to compensate victims of discharges, but the funds are only available if the discharger of the waste is clearly identifiable. The fund is used most often for compensating states for cleanup of spills from large tanker ships.

The Outer Continental Shelf (OCS) Lands Act sets up two funds to help pay for hazardous waste cleanup and to compensate victims. Under the act, OCS leaseholders are required to report spills and leaks from petroleum producing sites, and the Offshore Oil Spill Pollution Fund exists to finance cleanup costs and compensate injured parties for loss of the use of their property, loss of the use of natural resources, loss of profits, and a state or local government's loss of tax revenue. The U.S. Department of Transportation (DOT) is responsible for the administration of this fund with a balance maintained between $100 and $200 million provided by a 3 percent tax on oil produced at the OSC sites. If the operator of the site cooperates with DOT once a spill occurs, his or her liability is limited to a maximum of $35 million for a facility spill and $250,000 for a ship spill at the site.

The Fisherman's Contingency Fund also exists under the OCS Lands Act to repay fishermen for loss of profit and equipment due to oil and gas exploration, development, and production. The Secretary of the U.S. Department of Commerce is responsible for the administration of this fund. The balance of under $1 million is funded by assessments collected by the U.S. Department of the Interior from holders of permits and pipeline easements. If a fisherman cannot pinpoint the site responsible for the discharge of hazardous waste, his or her claim against the fund can still be acceptable if his or her boat was operated within the vicinity of OCS activity and if a claim was filed within 5 days of the injury. The fact that two agencies are involved in the administration makes this fund especially confusing to the victims of hazardous waste incidents.

The Price-Anderson Act was passed to provide compensation to victims of discharge from nuclear facilities. Nuclear incidents including radioactive material/toxic spills and emissions are covered as are explosions. The licensees of a nuclear facility are required by the Nuclear Regulatory Commission (NRC) to obtain insurance protection. If damages from an accident are greater than this insurance coverage, the federal government indemnifies the licensee up to $500 million. In no case does financial liability exceed $560 million. If a major meltdown had occurred at Three Mile Island in Pennsylvania, the inadequacy of this fund would have been painfully obvious.

Under the Deepwater Ports Act, the Coast Guard acts to remove oil from deepwater ports. The Department of Transportation administers a liability fund that helps pay for the cleanup and compensates injured parties. Financed by a $0.02 per barrel tax on oil loaded or unloaded at such a facility, the fund takes effect once the required insurance coverage is exceeded. The deepwater port itself must hold insurance for $50 million for claims against its waste discharges, and vessels using the port must hold insurance for $20 million for claims against their waste spills. Once these limits are exceeded, the fund established by the Deepwater Ports Act takes over. Again, the usefulness of the fund is dependent on how well two agencies work together and how well insurance claims are administered. This situation is again confusing at best to injured parties.

Statutes at the state level have generally paralleled these federal efforts. New Jersey has a Spill Compensation and Control Act for hazardous wastes listed by the EPA. The fund covers cleanup costs, losses of income, losses of tax revenues, and the restoration of damaged property and natural resources. A tax of $0.01 per barrel of hazardous substance finances a fund that pays injured parties if they file a claim within six years of the hazardous waste discharge and within one year of the discovery of the damage.

The New York Environmental Protection and Spill Compensation Fund is similar to this New Jersey fund, but with two major differences. In New York, only petroleum discharges are covered, and the generator may not blame a third party for a spill or discharge incident. Thus if a trucker or handler accidentally spills hazardous waste, the generator of the waste may still be held liable for damages.

Other states have limited efforts in the compensation of victims of hazardous waste incidents. Florida has a Coastal Protection Trust Fund of $35 million that compensates victims for spills, leaks, and dumping of waste. The coverage is limited, however, to injury due to oil, pesticides, ammonia, and chlorine, which provides a loophole for many types of hazardous waste discussed in Chapter 14. Other states that do have compensation funds generally spell out limited compensatable injuries and provide limited funds, often less than $100,000.

In the past, the value of these federal and state statutes has been questioned. Even when taken as a group, they do not provide for complete compensation strategy for dealing with hazardous waste. Few funds address personal injuries, and abandoned disposal sites are not considered. Huge administrative problems also exist as several agencies attempt to respond to a hazardous waste spill. Questions of which fund applies, which agency has jurisdiction, and what injuries can receive compensation linger.

To clean up their acts, the federal government enacted the Comprehensive Environmental Resource Conservation and Liability Act (CERCLA), commonly called the Superfund Act. This act evolved out of several years of battles between congressional committees, the EPA staff, and special interest groups. Originally, the Superfund Act was to serve two purposes: (1) provide money for the cleanup of abandoned hazardous waste disposal sites and (2) establish liability so dischargers could be made to pay for injuries and damages. Three considerations went into the development of this act.

- *What type of incidents should be covered?* The focus here was on decisions to include spills and abandoned sites under the Superfund Act as well as onsite toxic pollutants in harbors and rivers across the country. Fires and explosions caused by hazardous materials were also a focus of this consideration as the Superfund concept evolved through the years.
- *What type of damages should be compensated?* As alluded to above, three types of damages could be covered by a superfund: environmental cleanup costs; economic losses associated with property use, income, and tax revenues; and personal injury in the form of medical costs for acute injuries, chronic illness, death, and general pain and suffering.
- *Who pays for the compensation?* Choices here include federal appropriations into the fund, industry contributions from sales tax, or income tax surcharges. Federal and state cost sharing is also possible as are fees for disposal of hazardous waste at permitted disposal facilities.

The version adopted by Congress addressed each consideration and reflected the mood of compromise that surrounded the Superfund's evolution. A fund of between $1 billion and $4 billion was established that is financed 87 percent by a tax on the chemical industry and 13 percent by general revenues of the federal government. Small payments are permitted out of the fund, but only for out-of-pocket medical costs and partial payment for diagnostic services. The active Superfund Act does not set strict liability for spills and abandoned hazardous wastes.

Compensation for damages is one concern; regulations that control generators, transporters, and disposers of hazardous waste are the other side of the hazardous waste law coin. The Federal Resource Conservation and Recovery Act (RCRA), which is discussed at length in this chapter as it relates to solid waste, also is the principal statute that deals with hazardous waste. RCRA requires the EPA to establish a comprehensive regulatory program to control hazardous waste. This solid and hazardous waste act offers a good example of the types of reporting and recordkeeping requirements that are mandated in federal environmental laws. The Clean Water Act and Clean Air Act place requirements on water and air polluters similar to the types of requirements discussed below for generators of hazardous waste.

A generator of hazardous waste must meet these EPA requirements:

1. determine if the waste is hazardous as defined by an EPA listing of hazardous waste and by EPA testing procedures or as indicated by the materials and processes used in production
2. obtain an EPA general identification number
3. obtain a facility permit if hazardous waste is stored at the generating site for 90 days or longer
4. use appropriate containers and labels prior to shipment offsite
5. prepare a manifest for tracking the waste shipment offsite as described in Chapter 14
6. assure arrival of the waste at the disposal site
7. submit annual summaries of activities to federal and/or state regulatory agencies

The key to the regulatory system is this manifest system of "cradle-to-grave" tracking.

A transporter of hazardous waste must meet several requirements under RCRA:

1. obtain an EPA Transporter Identification Number
2. comply with the provisions of the manifest system
3. deliver the entire quantity of waste to the disposal/processing site
4. keep a copy of the manifest for 3 years
5. follow DOT rules for responding to spills of hazardous waste.

Again, the key element of transportation controls is the implementation and administration of the manifest system.

Facilities that treat, store, and/or dispose of hazardous waste also have requirements placed on them by RCRA.

1. The owner of the facility must apply for an operator's permit and supply data on the proposesd site and waste to be handled. This permit will spell out the terms of compliance: construction and operating schedules as well as monitoring and recordkeeping procedures.
2. As permits are granted by the federal or state agency, minimum operating standards will be placed on the facility. Design and engineering standards for containing, neutralizing, and destroying the wastes are addressed in these permits as are safety and emergency measures in the event of accident. Personnel training is included in each permit's requirements.

Table 16–1.   Checklist to Comply with Tosca's Premanufacture
Notification (PMN) System

Step	Requirement
1	Inspect EPA inventory of toxic substances to see if new product is listed.
2	Assess EPA rules for testing procedures to determine if substances not listed by the EPA are, in fact, "toxic."
3	Prepare and submit the PMN form at least 90 days prior to manufacture.
4	Obtain an EPA ruling: • *No ruling:* begin manufacturing at end of 90-day period • *EPA order:* respond to EPA order for more data • *EPA ruling:* EPA may prohibit or limit the proposed manufacturing, processing, use, or disposal of the substance.

3. The financial responsibility of the facility is defined, and a trust fund generally must be established at site closure. This fund enables the monitoring of groundwater and surface discharges to continue after the site is closed and enables the executor of the trust to maintain the site for years to come.

The key to the successful regulation of hazardous waste generation, transportation, and disposal is the manifest system and the manner in which the disposal processing facilities are regulated. The responsibilities for administering RCRA are passed from the EPA to state agencies as these agencies show they have the authority and expertise to effectively regulate hazardous waste.

In addition to the control of hazardous waste, Congress has passed the Toxic Substances Control Act (TOSCA) in an attempt to control hazardous substances before they become hazardous waste. One facet of this regulatory program is known as the Premanufacturing Notification (PMN) System. Under this system, all manufacturers must notify the EPA prior to marketing any substance not included in the EPA's 1979 inventory of toxic substances. In this notification, the manufacturer must analyze the predicted effect of the substance on workers, on the environment, and on consumers. This analysis must be based on test data and all relevant literature. A Compliance Checklist is presented in Table 16–1 where specific steps are spelled out for manufacturers who wish to produce a new product.

CONCLUSION

Solid waste statutory law developed rather rapidly once scientists and engineers realized the real and potential impacts of improper disposal.

Air, land, and water quality from such disposal became a key concern to local health officials and federal and state regulators. Hazardous and nonhazardous solid wastes are now regulated under a complex system of federal and state statutes that place operating requirements on facility operators.

PROBLEMS

16.1 Local land use ordinances often play key roles in limiting the number of sites available for a solid or hazardous waste disposal facility. Discuss the types of these zoning restrictions that apply in your home town.

16.2 Assume you work for a firm that contracts to clean the inside and outside of factories and office buildings in the state capitol. Your boss thinks he or she can make more money by expanding business to include the handling and transportation of hazardous wastes generated by current clients. Outline the types of data you must collect in order to advise him or her to expand/not expand.

16.3 Laws that govern refuse collection often begin with controls on the generator; that is, rules that each household must follow if city trucks are to pick up their solid waste. If you were asked by a town council to develop a set of such "household rules," what controls would you include? Emphasize public health concerns, minimizing the cost of collection, and even resource recovery considerations.

16.4 The oceans have long been viewed as bottomless pits into which the solid wastes of the world may be dumped. Engineers, scientists, and politicians are divided on the issue. Should ocean disposal be banned? Develop arguments pro and con.

LIST OF SYMBOLS

DOT = U.S. Department of Transportation
EPA = U.S. Environmental Protection Agency
NRC = Nuclear Regulatory Commission
OCS = outer continental shelf
PMN = Premanufacture Notification
RCRA = Resource Conservation and Recovery Act
TOSCA = Toxic Substance Control Act

Chapter 17

Air Pollution

Be it known to all within the sound
of my voice, whosoever shall be
found guilty of burning coal shall
suffer the loss of his head.
King Edward II, ca. 1300

Obviously, air pollution is not a new problem. King Edward II (1284–1327) tried to solve the problem of what Eleanor of Aquitaine called "the unendurable smoke" by prohibiting the burning of coal while Parliament was in session. His successors, Richard III (1377–1422) and Henry V (1413–1422), both took action against smoke, the former by taxation and the latter by forming a commission to study the problem. This was the first of a plethora of commissions, none of which helped reduce the level of air pollution in England. Under Charles II (1630–1685) a pamphlet was authored in 1661 by John Evelyn, entitled "Fumifugium or the Inconvenience of Aer and Smoak of London Dissipated, together with some Remedies Humbly Proposed." His suggestions included moving industry to the outskirts of town and establishing green belts around the city. None of his proposals were implemented. A subsequent commission in 1845 suggested, among other solutions to the smoke problems, that locomotives "consume their own smoke." In 1847 this requirement was extended to chimneys, and in 1853 it was decreed that offending chimneys be torn down.

In fact, for all the rhetoric and commissions, there was little action in England or anywhere else in the industrially developing world until after World War II. Action was finally prompted in part by two major air pollution episodes where human deaths were directly attributed to high levels of pollution.

The first of these occurred in Donora, a small steel town (pop. 14,000)

241

in western Pennsylvania. Donora is located in a bend of the Monongahela River and in 1948 had three main industrial plants—a steel mill, a wire mill, and a zinc plating plant.

During the last week of October 1948, a heavy smog settled in the area, and a weather inversion prevented the movement of pollutants out of the valley. (See Chapter 18 for a definition of inversion.) On Wednesday, the smog became especially intense. It was reported that streamers of carbon appeared to hang motionless in the air, and visibility was so poor that even natives of the area became lost.[1] By Friday, the doctor's offices and hospitals were flooded with calls for medical help. Yet no alarm had been sounded. The Friday Halloween parade was well attended, and a large crowd watched the Saturday afternoon football game between Donora and Monongahela High Schools.[2]

The first death had occurred, however, at 2 a.m. More followed in quick succession, and by midnight 17 persons were dead. Four more died before the effects of the smog abated. By this time the emergency had been recognized and special medical help was rushed in.

Although the Donora episode helped focus attention on air pollution problems in the United States, it took four more years before England suffered a similar disaster and action was finally taken. The "Killer Smog" of 1952 occurred in London with meteorological conditions similar to those during the Donora episode. A dense fog at ground level coupled with bitter cold and the smoke from coal burners caused the formation of another infamous "pea souper." This one was more severe than the usual smog however and lasted for over a week. The smog was so heavy that visibility during daylight hours was cut to only a few meters. Bus conductors had to walk in front of their vehicles to guide them through the streets. Two days after the fog set in, the death rate in London began to soar. Sulfur dioxide concentrations increased to nearly seven times their normal levels and carbon monoxide was twice the norm. It is important, however, to not conclude from the apparent relationship in Figure 17–1 that sulfur dioxides *caused* the deaths. Many other factors may have been responsible, and the various air pollutants probably acted synergistically to affect the death rate.

Primarily because of these acute episodes in London and Donora, public opinion and concern forced the initial attempts to clean up urban air.

---

[1] H. H. Schrenk, et al., "Air Pollution in Donora, Pa." U.S. Public Health Service Bulletin, No 306 (1949).

[2] It has been reported that the Monongahela coach protested the game and asked that it not be recorded in the books because he alleged that the Donora coach contrived to have a pall of smog hanging over the field so that, on kicking and passing plays, the ball would disappear and his players did not know where it would reappear.

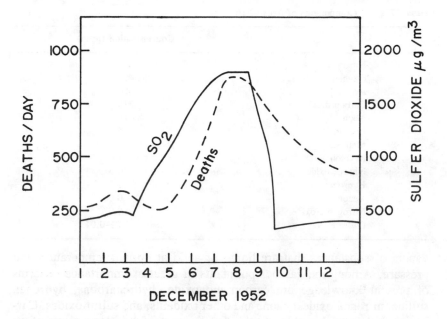

Figure 17-1 Deaths and pollutant concentration during the London 1952 air pollution episode. [*Source:* E. T. Wilkins, *Journal of the Royal San. Inst.* 74, 1 (1954).]

Interestingly, we have difficulty in defining what *is* clean air. From the scientific standpoint, clean air is composed of the constituents listed in Table 17-1. If we accept this as a definition of clean air, however, we are in trouble since any naturally occurring suspended material can be called a pollutant, and one never finds such "clean air" in nature. It may thus be more appropriate to define air pollutants as those substances that exist in such concentrations as to cause an unwanted effect. These pollutants can be natural (the blue haze over the Blue Ridge Mountains) or anthropogenic—"people-made"—(automobile exhaust). Air pollutants can be in the form of gases or suspended particulate matter (liquid or solid particles greater than 1μm in diameter).

TYPES OF AIR POLLUTANTS

Gaseous Pollutants

In the context of air pollution control, gaseous pollutants include substances that are gases at normal temperature and pressure as well as

Table 17–1.   Components of Normal Dry Air

	Concentration (ppm)
Nitrogen	780,900
Oxygen	209,400
Argon	9,300
Carbon dioxide	315
Neon	18
Helium	5.2
Methane	1.0–1.2
Krypton	1.0
Nitrous oxide	0.5
Hydrogen	0.5
Xenon	0.008
Nitrogen dioxide	0.02
Ozone	0.01–0.04

vapors of substances that are liquid or solid at normal temperature and pressure. Among the gaseous pollutants of greatest importance in terms of present knowledge are carbon monoxide, hydrocarbons, hydrogen sulfide, nitrogen oxides, ozone and other oxidants, and sulfur oxides. Carbon dioxide may be added to this list because of its potential effect on climate. These and other gaseous air pollutants are listed in Table 17–2.

Pollutant concentrations are commonly expressed as micrograms per cubic meter ($\mu g/m^3$). An older and still used method is to express gaseous pollutant concentrations as parts per million, where

$$1\ ppm = \frac{1\ volume\ of\ gaseous\ pollutant}{10^6\ volumes,\ pollutant\ +\ air}$$

and 1 ppm = 0.0001 percent by volume

Conversion between $\mu g/m^3$ and ppm is done using the ideal gas law

$$PV = nRT$$

where   P = pressure of gas
V = volume of gas
n = number of moles (weight/mol. wt.)
R = gas constant
T = temperature of gas in °K

Example 17.1
The exhaust of an automobile is found to contain 2 percent CO, at a temperature of 80 °C.

Express the concentration of CO in the exhaust in $\mu g/m^3$.

$$2\% = 20,000 \text{ ppm} = \frac{20,000 \text{ L}_{CO}}{10^6 \text{ L}_{exhaust}}$$

$T = 273° + 80° = 353°K$
$P = 1$ atmosphere
$R = 0.082$ L atm/mole $°K$

molecular weight of CO $= 12 + 16 = 28$ g/mole

From the ideal gas law

$$PV = nRT = \left(\frac{\text{weight of CO}}{\text{mol. wt.}} \right) RT$$

Or, solving for the weight of CO:

$$\text{weight of CO} = \frac{PV(\text{mol. wt.})}{RT} = \frac{(1)(20,000)(28)}{(0.082)(353)} = 19,346 \text{ g}$$

Thus

$$20,000 \text{ L}_{CO}/10^6 \text{ L}_{exhaust} = \frac{1.93 \times 10^4 g}{10^6 L}$$

$$= \frac{1.93 \times 10^4 g}{10^3 m^3}$$

$$= 19300 \mu g/m^3$$

Particulate Pollutants

Two air pollutants of national concern exist as particulates in the air. One of these is lead; the other is the general class of particulates—chemistry unspecified.

Particulate pollutants are classified as follows:

1. *Dust*—solid particles that are (a) entrained by process gases directly from the material being handled or processed, for example, coal, ash, and cement; (b) direct offspring of a parent material undergoing a mechanical operation, for example, sawdust from woodworking; (c) entrained mate-

Table 17–2.   Gaseous Air Pollutants

Name	Formula	Properties of Importance	Significance as Air Pollutant
Sulfur dioxide	SO_2	Colorless gas, intense choking odor, highly soluble in water to form sulfurous acid H_2SO_3	Damage to vegetation, property, and health
Sulfur trioxide	SO_3	Soluble in water to form sulfuric acid H_2SO_4	Highly corrosive
Hydrogen sulfide	H_2S	Rotten egg odor at low concentrations, odorless at high concentrations	Highly poisonous
Nitrous oxide	N_2O	Colorless gas, used as carrier gas in aerosol bottles	Relatively inert. Not produced in combustion
Nitric oxide	NO	Colorless gas	Produced during high-temperature, high-pressure combustion. Oxidizes to NO_2
Nitrogen dioxide	NO_2	Brown to orange gas	Major component in the formation of photochemical smog
Carbon monoxide	CO	Colorless and odorless	Product of incomplete combustion. Poisonous
Carbon dioxide	CO_2	Colorless and odorless	Formed during complete combustion. Possible effects in producing changes in global climate
Ozone	O_3	Highly reactive	Damage to vegetation and property. Produced mainly during the formation of photochemical smog
Hydrocarbons	C_aH_y or HC	Many	Some hydrocarbons are emitted from automobiles and industries, others are formed in the atmosphere
Hydrogen fluoride	HF	Colorless, acrid gas	Product of aluminum refining, causes fluorosis in cattle.

rials used in a mechanical operation, for example, sand from sandblasting. Dusts from grain elevators and coal-cleaning plants typify this class of particulate. Dust consists of relatively large particles. Cement dust, for example, is about 100 μ in diameter.

2.   *Fume*—a solid particle, frequently a metallic oxide, formed by the condensation of vapors by sublimation, distillation, calcination, or chemical reaction processes. Examples of fumes are zinc and lead oxide resulting from

the condensation and oxidation of metal volatilized in a high-temperature process. The particles in fumes are quite small with diameters from 0.03 to 0.3 μ.

3. *Mist*—a liquid particle formed by the condensation of a vapor and perhaps by chemical reaction. An illustration of this process is the formation of sulfuric acid mist:

$$SO_3 \text{ (gas) } 22\,°C \rightarrow SO_3 \text{ (liquid)}$$

$$SO_3 \text{ (liquid) } + H_2O \rightarrow H_2SO_4$$

Sulfur trioxide gas becomes a liquid since its dew point is 22 °C and SO_3 particles are hydroscopic. Mists typically range from 0.5 to 3.0 μ in diameter.

4. *Smoke*—solid particles formed as a result of incomplete combustion of carbonaceous materials. Although hydrocarbons, organic acids, sulfur oxides, and nitrogen oxides are also produced in combustion processes, only the solid particles resulting from the incomplete combustion of carbonaceous materials are smoke. Smoke particles have diameters from 0.05 to approximately 1 μ.

5. *Spray*—a liquid particle formed by the atomization of a parent liquid.

Approximate size ranges of the various types of air pollutants are shown in Figure 17–2. The concentration of particulates is always expressed as $\mu g/m^3$.

Particulate air pollutants made up of particles smaller than 10 μm in diameter ("respirable particulate matter") are of particular concern as are especially hazardous air pollutants like asbestos, heavy metal compounds, radioactive substances, and certain organic compounds.

The Greenhouse Effect

Carbon dioxide and water vapor are also constituents of the atmosphere but are not presently considered pollutants. There is mounting evidence that the atmosphere is accumulating an excess of CO_2 sufficient to bring about some climate changes. Internal vibration and rotation of molecular CO_2 cause it to absorb infrared radiation. Carbon dioxide in the air will thus absorb some of the heat the earth normally radiates into space and reradiate that heat back to the earth's surface. A canopy of CO_2 has an effect similar to glass in a greenhouse: It traps heat that would otherwise be lost by radiation. It is possible that an accumulation of atmospheric CO_2 could trap enough heat to warm the earth considerably and eventually cause some melting of the polar ice caps. Evidence is accumulating that the earth may be getting warmer. Other gases which absorb in the

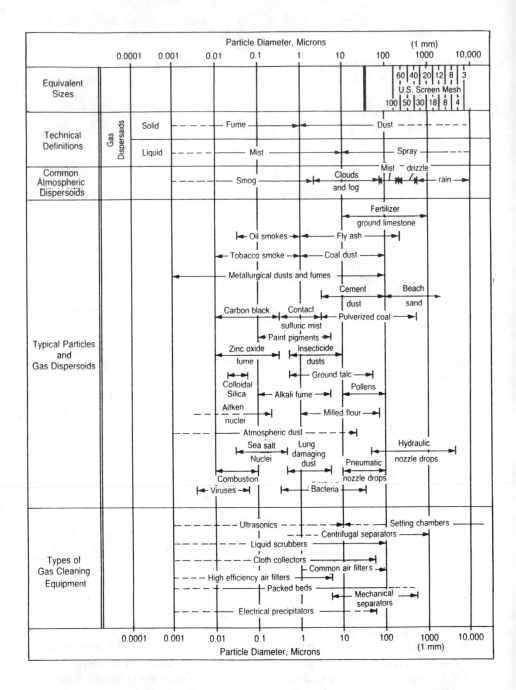

Figure 17-2   Characteristics of particles and particle dispersolds. [*Source: Adapted from C.E. Lapple. Stanford Research Institute Journal* 5   (1961).]

infrared can enhance the effect of CO_2. All these gases are sometimes referred to collectively as "greenhouse gases."

SOURCES OF AIR POLLUTION

Many of the pollutants of concern are formed and emitted through natural processes. For example, naturally occurring particulates include pollen grains, fungus spores, salt spray, smoke particles from forest fires, and dust from volcanic eruptions. Gaseous pollutants from natural sources include carbon monoxide as a product of animal and plant respiration, hydocarbons (terpenes) from conifers, sulfur dioxide from geysers, hydrogen sulfide from the breakdown of cysteine, and other sulfur-containing amino acids by bacterial action, nitrogen oxides, and methane.

Near cities and in populated areas more than 90 percent of the volume of air pollutants is the result of human activity. Anthropogenic sources of air pollution may be broadly classified as stationary or mobile sources. Stationary sources include combustion processes, solid waste disposal, industrial processes, construction, and demolition. The principal pollutants from stationary combustion processes are particulate matter, in the form of fly ash and smoke, and the gaseous pollutants sulfur dioxide, nitric oxide, and nitrogen dioxide.

Sulfur Dioxide

For uncontrolled sources, sulfur dioxide emissions are a function of the amount of sulfur present in the fuel. Thus combustion of fossil fuels that contain sulfur can yield significant quantities of SO_2.

In the United States, the largest single source of atmospheric sulfur dioxide is fossil fuel combustion both for electric power generation and for process heat. In any given community, however, an industrial plant may be the principal source since many industrial operations emit significant quantities of SO_2 as well. Some significant industrial emitters are:

- *Nonferrous smelters.* With the exception of iron and aluminum, metal ores are sulfur compounds. When the ore is reduced to the pure metal, the sulfur in the ore is ultimately oxidized to SO_2. Copper ore, for example, is usually a form of CuS. Since the atomic weight of copper is 64 gm/mole—the same as the molecular weight of SO_2—every pound of copper refined yields a pound of sulfur dioxide, which goes into the air unless it is trapped.
- *Oil refining.* Sulfur and hydrogen sulfide are constituents of crude oil, and H_2S is released as a gas during catalytic cracking. Since hydrogen sulfide is considerably more toxic than sulfur dioxide, it is flared to SO_2 before release to the ambient air.

- *Pulp and paper manufacture.* The sulfite process for wood pulping uses hot H_2SO_3 and thus emits SO_2 into the air. The kraft pulping process produces H_2S, which is then flared to SO_2.

Oxides of Nitrogen

Nitric oxide is formed by the thermal fixation of atmospheric nitrogen. The reaction is strongly temperature dependent. The equilibrium constant for the reaction

$$N_2 + O_2 = 2NO$$

is approximately proportional to the fourth power of the absolute temperature at which the reaction takes place. Thus, all high-temperature processes produce NO, which is then oxidized further to NO_2 in the ambient air.

In the United States, about half the atmospheric nitrogen oxides are produced by stationary sources and about half, by mobile sources. Fossil fuels contain nitrogen compounds, which are also oxidized to NO on combustion.

Carbon Monoxide

CO is a product of incomplete combustion of carbon-containing compounds. Stationary combustion sources produce CO, which is usually transformed to CO_2 rapidly enough so that there is little significant dispersion of CO from the stationary source in the ambient air.

Most of the CO in the ambient air comes from vehicle exhaust. Internal combustion engines do not burn fuel completely to CO_2 and water; some unburned fuel will always be exhausted with CO as a component. Various afterburner devices can be used to reduce CO.

CO is a localized pollutant and tends to build up in areas of concentrated vehicle traffic, in parking garages, and under building overhangs.

Hydrocarbons

Vehicles are also a major source of atmospheric hydrocarbons. Stationary sources of hydrocarbons include petrochemical manufacture, oil refining, incineration, paint manufacture and use, and dry cleaning.

Ozone (Photochemical Oxidant)

Photochemically formed organic oxidants, which are classified as ozone, are a secondary air pollutant. That is, ozone is not emitted directly into the air but is the result of chemical reactions in the ambient air. The components of automobile exhaust are particularly important in the formation of atmospheric ozone and are the chief contributors to Los Angeles smog, which is mostly ozone. Table 17–3 lists in simplified form some of the key reactions in the formation of photochemical smog.

Table 17–3.   Simplified Reaction Scheme for Photochemical Smog

NO_2 Nitrogen Dioxide	+ Light	→ NO Nitric oxide	+ O Atomic oxygen
O	+ O_2 Molecular oxygen	→O_3 Ozone	
O_3	+ NO	→NO_2	+ O_2
O	+ HC Hydrocarbon	→HCO^* Radical	
HCO^*	+ O_2	→ HCO_3^* Radical	
HCO_3^*	+ HC	→ Aldehydes, ketones, etc.	
HCO_3^*	+ NO	→HCO_3^* Radical	+ NO_2
HCO_3^*	+ O_2	→ O_3	+ HCO_3^*
HCO_x^* Radical	+ NO_2	→ Peroxyacetyl nitrates	

The reaction sequence illustrates how nitrogen oxides formed in the combustion of gasoline and other fuels and emitted to the atmosphere are acted upon by sunlight to yield ozone (O_3), a compound not emitted as such from a source and hence considered a secondary pollutant. Ozone in turn reacts with hydrocarbons to form a series of compounds that includes aldehydes, organic acids, and epoxy compounds. Thus the atmosphere can be viewed as a huge reaction vessel wherein new compounds are being formed while others are being destroyed.

The formation of photochemical smog is a dynamic process. Figure 17–3 is an illustration of how the concentrations of some of the components vary during the day. Note that as the morning rush hour begins the NO levels increase, followed quickly by NO_2. As the latter reacts with sunlight, O_3 and other oxidants are produced. The hydrocarbon level similarly increases at the beginning of the day and then drops off in the evening.

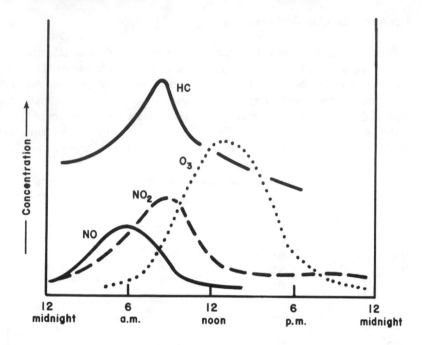

Figure 17-3 Formation of photochemical smog.

The reactions involved in photochemical smog were a mystery. Particularly baffling was the formation of high ozone levels. As seen from the first three reactions in Table 17-3, for every mole of NO_2 reacting to make atomic oxygen and hence ozone, one mole of NO_2 was created from reaction with the ozone. All of these reactions are fast. How then could the ozone concentrations build to such high levels?

One answer is that NO enters into other reactions, especially with various hydrocarbon radicals, and thus allows excess ozone to accumulate in the atmosphere (the seventh reaction in Table 17-3). In addition, some hydrocarbon radicals react with molecular oxygen and also produce ozone.

Acid Rain

Acid precipitation is another pollutant formed by reactions in the air. Sulfur dioxide and nitrogen dioxide react with water and atmospheric oxygen to produce sulfuric and nitric acid. Normal, uncontaminated rain has a pH of about 5.6, but acid ran can be as low as pH 2.5. Hundreds of lakes in North America and Scandinavia have become so acidic that they no longer can support fish life. In a recent study of Norwegian lakes, more than 70 percent of the lakes having a pH of less than 4.5 contained no fish, and nearly all lakes with a pH of 5.5 and above contained fish. The low pH not only affects fish directly but also contributes to the release of potentially toxic metals such as aluminum, thus magnifying the problem. In Norway, storms that travel over the industrial areas of Great Britain and continental Europe can be tracked and have been found to dump especially destructive precipitation. The recognition of this problem makes the use of the "tall stack" method of air pollution control highly questionable (see Chapter 20).

In North America acidification has already wiped out all fish and many plants in 50 percent of the high mountain lakes in the Adirondacks. The pH in many of these lakes has reached such levels of acidity as to replace the trout and native plants with acid-tolerant mats of algae.

Suspended Particulate Matter

Virtually every industrial process is a potential source of dust, smoke, or aerosol emissions. Waste incineration, coal combustion, and the combustion of heavy (bunker grade) oil are sources. Agricultural operations are a major source of dust especially in dry land farming. Traffic on roads, even on completely paved roads, is also a major source of dust. In 1985, 40 percent of the total suspended particulate matter in the air of Seattle's downtown industrial center was identified as "crustal dust," dust raised by traffic on paved streets.[3] Construction and demolition can be particularly troublesome sources of dust in urban areas.

Fires are a major source of airborne particulate matter as well as of hydrocarbon emissions and carbon monoxide. Forest fires are usually considered a natural (nonanthropogenic) source, but fires for land clearing, slash burns, agricultural burns, and trash fires contribute considerably. Since 1970, most communities in the United States have prohibited open burning of trash and dead leaves.

Wood-burning stoves and fireplaces also produce smoke that contains partly burned hydrocarbons, aromatic compounds, tars, and aldehydes as well as smoke and ash. Several cities (the largest is Portland, Oregon) now prohibit the use of wood stoves and fireplaces during unfavorable weather conditions.

---

[3]Schoefield, C. Cornell University, as reported in the *Washington Post* (7 January 1981).

Lead

Airborne lead is particulate matter and is chemically generally lead oxide or lead chloride. Vehicle exhaust is virtually the only source of airborne lead except in the vicinity of nonferrous smelters. The Bunker Hill lead smelter at Kellogg, Idaho—the largest in the United States—was shut down in 1980, eliminating a significant source of lead pollution in Kellogg.

Lead from vehicle exhaust is deposited for several hundred yards downwind of highways. In urban areas where there is a concentration of freeway intersections, this deposited lead has been identified as a source of lead intoxication.

HAZARDOUS SUBSTANCES

There are a number of hazardous substances emitted into the air that pose particular control problems. Some of these, together with several sources that have been identified are:

- *Asbestos:* construction, demolition, and remodeling of existing structures, replacement of pipes, furnaces, asbestos refining and fabrication, dust from soil erosion, and so forth.
- *Mercury:* chlor-alkali manufacture, battery manufacture.
- *Vinyl chloride monomer:* manufacture and fabrication of polyvinylchloride products
- *Fluorides:* primary aluminum smelting, phosphate fertilizer manufacture.
- *Hydrogen sulfide:* crude oil transportation, oil desulfurizing and refining, kraft paper manufacture.
- *Benzene:* petrochemical manufacture, industrial solvent use, pharmaceutical manufacture.

HEALTH EFFECTS

Much of our knowledge of the effects of air pollution on people comes from the study of acute air pollution episodes. As previously noted, the two most famous episodes in the past occurred in Donora, Pennsylvania, and in London, England. In both episodes the illness appeared to be chemical irritation of the respiratory tract. The weather circumstances were also similar in that a high pressure system with an inversion layer was present. An inversion, which is discussed further in Chapter 18, is a layer of warm air aloft that prevents the pollutants from escaping and diluting vertically. In addition, inversions are usually accompanied by low surface winds leading to reduced horizontal dilution of pollutants.

In both episodes, the pollutants affected a specific segment of the public—those individuals already suffering from diseases of the cardio-respiratory system. Another observation of great importance is that it was not possible to blame the adverse effects on any one pollutant. This observation puzzled the investigators (industrial hygiene experts) who were accustomed to studying industrial problems where one could usually relate health effects to a specific pollutant. Today, after many years of study, it is thought that the health problems during the episodes were attributable to the combined action of a particulate matter (solid or liquid particles) and sulfur dioxide, a gas. No one pollutant by itself, however, could have been responsible.

A major catastrophic air pollution episode was recorded in 1984 with the release of methyl isocyanate into the air in Bhopal, India. More than 1500 deaths occurred among residents of the area within four miles of the plant, and thousands of people suffered temporary or permanent skin burns, eye damage, and damage to the respiratory tract and nervous system.

Until recently scientists had only such episodes from which to evaluate the health effects of air pollution. Laboratory studies with animals are of some help, but the step from a rat to a man (anatomically speaking) is quite large. There are a number of long-term studies of the health effects of air pollution now under way in large U.S. urban centers.

The major target of air pollutants is the respiratory system, pictured in Figure 17–4. Air (and entrained pollutants) enter the body through the throat and nasal cavities and pass to the lungs through the trachea. In the lungs, the air moves through bronchial tubes to the alveoli, small air sacks in which the gas transfer takes place. Pollutants are either absorbed into the bloodstream or moved out of the lungs by tiny hair cells called *cilia* that are continually sweeping mucus up into the throat. The bronchial cilia can be effectively paralyzed by inhaled smoke, enhancing the *synergistic effects* between smoking and air pollutants. Pollutant particles small enough to pass the bronchial cilia by are called *respirable particles*.

Carbon Monoxide

The effect of carbon monoxide inhalation on human health is directly proportional to the quantity of CO bound to hemoglobin. These effects are summarized in Table 17–4.

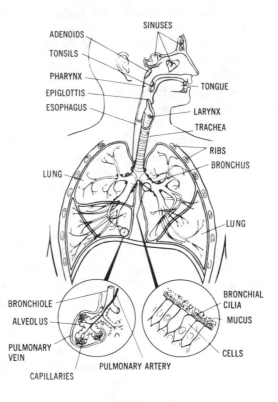

Figure 17–4   The respiratory system (courtesy American Lung Association).

Oxygen is transported in the blood as oxyhemoglobin (HbO_2), a semistable compound in which O_2 is weakly bound to the Fe^{2+} in hemoglobin in red blood cells. The O_2 is removed for cell respiration, and the regenerated hemoglobin is available for more oxygen transport. CO reduces the oxygen-carrying capacity of the blood by combining with hemoglobin and forming carboxyhemoglobin (HbCO), which is stable. Hemoglobin tied up as HbCO cannot be regenerated and is not available for oxygen transport for the life of that particular red blood cell. CO effectively poisons the hemoglobin oxygen transport system.

Hemoglobin has a greater affinity for carbon monoxide than it does for oxygen. A person who breathes a mixture of CO and oxygen will carry equilibrium concentrations of HbCO and HbO_2 given by:

$$[HbCO]/[HbO_2] = \frac{M\ p(CO)}{p(O_2)}$$

Table 17–4.   Health Effects of Carbon Monoxide and COHb

Carbon Monoxide

Environmental Condition	Effects
9 ppm 8-hr exposure	Ambient air quality standard
50 ppm 6-wk exposure	Structural changes in heart and brain of animals
50 ppm 50-min exposure	Changes in relative brightness threshold and visual acuity
50 ppm 8 to 12-hr exposure nonsmokers	Impaired performance on psychomotor tests

Carboxyhemoglobin

COHb Level (percent)	Effects
<1.0	No apparent effect
1.0–2.0	Some evidence of effect on behavioral performance
2.0–5.0	Central nervous system effects. Impairment of time-interval discrimination, visual acuity, brightness discrimination, and certain other psychomotor functions
>5.0	Cardiac and pulmonary functional changes
10.0–80.0	Headaches, fatigue, drowsiness, coma, respiratory failure, death

where $p(CO)$ = partial pressure of CO
$p(O_2)$ = partial pressure of O_2
M = a constant whose range in human blood is from 200 to 250

Example 17.2

Make an estimate of the saturation value of HbCO in the blood if the CO content of the air breathed is 100 ppm.

Assume a value for M of 210. Then since air is 21% oxygen,

$$[HbCO]/[HbO_2] = M\, p(CO)/p(O_2)$$
$$= (210)(100)/210,000$$
$$= 0.1$$

Thus $[HbCO]/[HbO_2] = 0.1$ or 10%

HbCO levels as a function of exposure are shown in Figure 17–5.

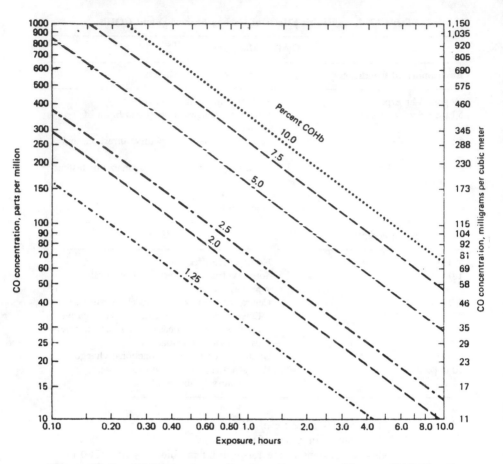

Figure 17–5   HbCO level as a function of exposure. [*Source:* NAPCA. *Air Quality Criteria for Carbon Monoxide,* AP-62, Washington, D.C.: HEW, 1970.]

Particulate Matter

The site and the extent of the deposition of particulates in the respiratory tract is a function of certain physical factors, the most important one being particle size.

Respirable particulate matter is especially harmful because it is small enough to penetrate the respiratory system down to the pulmonary alveoli. Alveolar deposition is especially important since that region of the lungs is not provided with cilia to remove particulates. Thus particles deposited there remain for a relatively greater length of time.

The cilia (Figure 17–4), which protect the respiratory system by sweeping out particles, are also affected by sulfur dioxide. Experiments using rabbits and other animals have shown that the frequency with which cilia beat is decreased in the presence of sulfur dioxide. Thus, sulfur dioxide, in addition to constricting the bronchi, also affects the protection mechanism of the respiratory tract.

Nitrogen dioxide (NO_2) is a pulmonary irritant. Although little is known of the specific toxic mechanism, it is known that nitrogen dioxide at concentrations admittedly greater than those found in community air is an edema producer and also results in pulmonary hemorrhage. Edema is an abnormal accumulation of fluid in body tissues. Pulmonary edema is excessive fluid in the lung tissues.

Nitric oxide (NO), the other common oxide of nitrogen, is not an irritant gas. *In vitro* nitric oxide combines readily with human hemoglobin to form a highly stable nitric oxide hemoglobin, but interestingly this hemoglobin effect has not been observed in living animals.

Ozone (O_3) is a highly irritating, oxidizing gas. Concentrations of a few parts per million produce pulmonary congestion, edema, and hemorrhage. A one-hour exposure of human subjects to 2500 $\mu g/m^3$ can increase the residual lung volume and decrease maximum breathing capacity. The symptoms of ozone exposure are initially a dry throat followed by headache, disorientation, and altered breathing patterns.

A fascinating observation from animal studies is the development of ozone tolerance. An animal exposed initially to a low concentration of ozone can survive a subsequent exposure to an ordinarily fatal concentration of ozone. To what extent ozone tolerance develops in humans is unknown.

Ozone in the air we breathe should not be confused with the so-called "ozone layer" in the stratosphere. The stratosphere begins at an altitude of seven to ten miles, depending on the latitude and season of the year. Ozone in this thin air absorbs a large part of the sun's ultraviolet radiation. It is believed that vapor from supersonic aircraft flights and fluorocarbon gases from spray cans have caused reduction in stratospheric ozone that may be permanent. This can increase the ultraviolet radiation reaching the earth, raising the incidence of human skin cancer and probably affecting the earth's climate and ecological systems in unpredictable ways. The high ozone layer seems to be maintained by natural processes, mainly the sun's radiation. Lightning discharges are the principal natural source of ozone in the lower atmosphere. The recently discovered polar holes in the ozone layer are of serious concern.

Among the organic gases, formaldehyde (HCHO) is particularly important. Its effects are similar to those of sulfur dioxide—that is, irritation of mucous membranes and bronchial constriction.

It has also been noted that a linear statistical association exists between aldehyde concentrations and eye irritation. An unsaturated aldehyde, acrolein (CH_2CHCHO), at concentrations of a tenth of a part per million will produce eye irritation comparable to that which occurs in Los Angeles photochemical smog. Acrolein is also irritating to mucous membranes and produces bronchoconstriction. Some of the chronic diseases believed to be causally related to air pollution are lung cancer and chronic obstructive pulmonary disease (COPD of which emphysema is one manifestation). Asthma and other respiratory allergies appear to be aggravated by air pollution. The following facts have been cited in support of the hypothesis that lung cancer is related to air pollution: (1) lung cancer mortality is higher in very large cities than it is in rural areas, even when cigarette smoking is taken into account; (2) substances that cause cancer in experimental animals (both organic and metallic) are found in polluted atmospheres; (3) some carcinogens are relatively stable in the atmosphere; (4) compounds extracted from air samples produce cancers in bioassay animals; (5) atmospheric irritants affect the protective action of cilia and mucous flow; (6) high workplace concentrations of some air pollutants have produced specific types of lung cancer in exposed workers; and (7) both the carcinogens and irritants have physical characteristics compatible with the postulated effects. Even this statistical and experimental evidence, however, does not prove conclusively that outdoor air pollution *does* cause cancer. Indoor air pollution, which can be severe in densely populated cities, may contribute. The actual cause-and-effect relationship is still a medical mystery.

Chronic bronchitis is a disorder characterized by excessive mucus secretion in the bronchial tubes. It is manifested by a chronic or recurrent productive cough. In Great Britain, chronic bronchitis is the second most common cause of death in men aged 40 to 55 and is the leading cause of disability in this age group.

Emphysema is the breakdown and destruction of the alveolar walls in the lung. The destruction of these alveolar sacs leads to great difficulty in breathing. Available evidence suggests a relationship between emphysema and polluted air. In the United States, for example, the disease is more common in urban than in nonurban areas and is a leading cause of death, especially for cigarette smokers.[5]

In addition to the chronic diseases, air pollution can have a serious influence on acute diseases such as the common cold and pneumonia. Statistical evidence has shown that both of these illnesses can be aggravated by breathing dirty air.

---

[5] The tough, weathered cowboy who appears in the Marlboro cigarette advertisements recently died of pneumonia aggravated by emphysema.

EFFECTS ON VEGETATION

Vegetation is injured by air pollutants in three ways: (1) necrosis (collapse of the leaf tissue), (2) chlorosis (bleaching or other color changes), and (3) alterations in growth. The types of injury caused by various pollutants differ markedly (Figure 17–6).

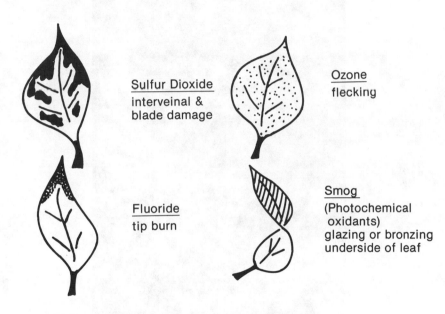

Figure 17–6 Typical air pollutant injury to vegetation.

Sulfur dioxide produces marginal or interveinal blotches that are white to straw in color on broad-leafed plants (Figure 17–7). Grasses injured by sulfur dioxide show a streaking (light tan to white) on either side of the midvein. Brown necrosis occurs on the tips of conifer needles with adjacent chlorotic areas. Alfalfa, barley, cotton, wheat, and apple are among the plants most sensitive to sulfur dioxide. Sensitive species are injured by exposure to concentrations of 780 $\mu g/m^3$ for 8 hours.

In conifers and grasses, fluoride exposure produces an injury known as tip burn. In broad-leafed plants the fluoride injury is a necrosis at the periphery of the leaf. Among the plants most sensitive to fluorides are gladiolus, Chinese apricot, Italian prune, and pine.

At sufficient concentrations ozone produces tissue collapse and markings of the upper surface of the leaf. These markings are known as stipple (pigmented red-brown) and flecking (bleached straw to white). One to two

Figure 17-7   Sulfur dioxide injury to oak.

hours of exposure at air levels of about 300 $\mu g/m^3$ produces injury in sensitive species. Sensitive varieties include tomato, tobacco, bean, spinach, and potato.

Peroxyacyl nitrates (PAN) are present in photochemical smog and produce typical smog injury. The smog-produced injury is a bronzing on the underside of the leaves of vegetables. In grasses, the collapsed tissue shows up as bands bleached tan to yellow. Smog exposure has also been shown to produce early maturity or senescence in plants. Even low PAN concentrations will injure sensitive species. Among the most sensitive species are petunia, romain lettuce, pinto bean, and annual bluegrass.

EFFECTS ON DOMESTIC ANIMALS

As might be expected, air pollutants affect animals other than people. During the Donora episode, 20 percent of the canaries and 15 percent of the dogs were affected.

In the 1952 London episode, of 351 cattle on the ground floor at the Smithfield Cattle Show, 52 became seriously ill, 5 died, and 9 others were slaughtered later.

Chronic poisoning usually results from ingesting forage contaminated by the pollutant. Pollutants important in this connection are the heavy metals arsenic, lead, and molybdenum. At Anaconda, Montana, in 1902, extensive poisoning of cattle, horses, and sheep occurred in the vicinity of a copper smelter. Of a flock of 3500, about 620 sheep died after feeding on vegetation some 15 miles from the smelter. In 1954 cattle grazing at a distance of 0.1 km from a steel plant in Sweden were poisoned by molybdenum.

Perhaps the best known of the pollutants affecting livestock is fluoride. The problem of *fluorosis* of livestock is an old one. Farm animals, particularly cattle, sheep, and swine, are susceptible to fluorine poisoning. Deer in wildlife refugees downwind from fluoride emitters have contracted fluorosis. Horses and poultry, on the other hand, seem to have a high level of resistance. Fluorosis is characterized by mottled teeth and a condition of the joints known as exostosis leading to lameness and ultimately death.

EFFECTS ON MATERIALS

Perhaps the most familiar effect of air pollution on materials is soiling of building surfaces, clothing, and other articles. Soiling results from the deposition of smoke (fine particles of approximately 0.3 μ diameter) on surfaces. Over a period of time this deposition becomes noticeable as soiling, a discoloring, or darkening of the surface. Damage to the surface, of course, results from the cleaning operation. In the case of exterior building materials, sandblasting is often required to clean the surface—and part of the surface is removed in the process of cleaning.

Another effect of air pollution is that of accelerating the corrosion of metals. For example, it has been observed that in the presence of sulfur dioxide many materials corrode much faster than they would otherwise.

One of the early noted effects of the Los Angeles smog was rubber cracking. Indeed, the effect of ozone, a principal ingredient of smog, on rubber is so specific that rubber cracking can be used to measure ozone concentrations (Chapter 19). The economic significance of rubber cracking is apparent.

The action of hydrogen sulfide on lead-base paints is well known. Hydrogen sulfide, in the presence of moisture, reacts with lead dioxide in paint to form lead sulfide, producing a brown to black discoloration. As might be imagined, this is unattractive on a white house.

EFFECTS ON ATMOSPHERE

The ability of air pollutants, especially particulates, to reduce visibility is well known to passengers on airplanes approaching nearly any large urban area in the United States. The visibility reduction results from light scattering rather than obscuration of light. The particles primarily responsible for this effect are quite small—in the range of 0.3 to 0.6 μ in diameter.

Although carbon dioxide is usually not considered a pollutant, CO_2 concentrations in the atmosphere have been increasing in recent years. Although all oxidation of carbon compounds produces CO_2, the recent large increases are attributed to increased combustion of fossil fuels. Since CO_2 absorbs strongly at about 15 μ (heat energy) it retards the radiative cooling of the earth. The CO_2 concentrations will increase from the 1968 level of 0.032 percent to about 0.038 percent by the year 2000. This would theoretically result in an increase of 0.5 °C, and is often referred to as the greenhouse effect. Several other gases also absorb in this region, some aerosol propellants for instance. These are collectively referred to as "greenhouse gases."

In contrast to the above, some observations have indicated that the temperature near the earth's surface has declined in recent years. The investigators attribute this decline to an increase in the earth's albedo (reflectivity) as a result of an increased atmospheric aerosol from pollution. Measurements of atmospheric turbidity in Washington, D.C., and Davos, Switzerland, indicate increases of particulates of up to 57 and 70 percent respectively during the first half of this century. These observations support the suggestion of increased albedo or reflectivity. Is it possible that the aerosol effect is keeping pace with the CO_2 effect?

A second global atmospheric effect of air pollution is the destruction of the ozone layer due to the emission of chlorofluorocarbons (freons). The greatest source of these emissions is from chlorofluorocarbon aerosol propellants. Leaking refrigeration systems are also a source. Substitution of other fluids for chlorofluorocarbons should be a major global objective.

CONCLUSION

It is important to realize that while the effect of air pollutants on animals, vegetation, and materials is easy to determine, the health effects on humans can only be estimated on the basis of epidemiological evidence. Due to moral and ethical considerations that preclude deliberate exposure of human subjects to concentrations that might result in disease, this is the only avenue of study open. Epidemiological studies have been criticized because the results are expressed as statistical associations. It should be pointed out that it is probably the best evidence we will ever have. In addition, these studies are with people in the real world, exposed to real pollution, not to a synthetic atmosphere concocted in the laboratory.

One thing is certain—all of the evidence, both experimental and epidemiological, suggests that heavily polluted air can indeed be a serious threat to our health and well-being.

PROBLEMS

17.1 If you drive a 1974 car an average of 1000 miles/month, how much CO and HC would be emitted during the year? (Note: The EPA 1974 standards are 3.4 g/mi for HC and 39 g/mi of CO.) How long would it take to achieve lethal concentration of CO in a common doublecar garage?

17.2 A 2.5 percent level of CO in hemoglobin (HbCO) has been shown to cause impairment in time-interval discrimination. The level of CO on crowded city streets sometimes hits 100 ppm CO. An approximate relationship between CO and HbCO (after prolonged exposure) is

$$HbCO(\%) = 0.5 + 0.16 \times CO(ppm)$$

What level of HbCO would a traffic cop be subjected to during a working day directing traffic on a city street?

17.3 Draw a graph showing the concentration of NO, NO_2, HC, and O_3, in the Los Angeles area during a sunny smoggy day. Superimpose on this graph, in another color, the curves as they appear on a cloudy day.

17.4 Give three examples of synergism in air pollution.

17.5 If SO_2 is so soluble in water, how can it get to the deeper reaches of the lung without first dissolving in the mucus?

17.6 Match the pollutants with the appropriate description.

A.   Hydrogen fluoride (HF)   _____ a secondary pollutant in photo-chemical smog

B.   Hydrogen sulfide (H_2S)   _____ cause of "weather fleck"

C.   Carbon monoxide (CO)   _____ moved by cilia

D.   Carbon dioxide (CO_2)   _____ one major source is power production

E.   Sulfur dioxide (SO_2)   _____ combines with hemoglobin

F.   Particulates   _____ a pungent, toxic gas often found in sewers

G.   Oxides of nitrogen (NOx)   _____ result of phosphate fertilizer production

H.   Peroxyacyl nitrate (PAN)   _____ produced naturally by decaying matter

I.   Ozone (O_3)   _____ produced only by high compression combustion

17.7 The state of Florida has set an annual average ambient SO_2 standard of 0.003 ppm. At ordinary ambient temperatures, what would this be in $\mu g/m^3$?

17.8 The state of Michigan set ambient H_2S standards at 2 ppb. Assuming that the compound is diluted by a factor of 10,000 after emission from the stack and that the stack gas is at a temperature of 100°C, what is the allowed H_2S emission from a kraft pulp mill expressed in $\mu g/m^3$?

Chapter 18

Meteorology and Air Quality

One would think that, with the earth's atmosphere being about 100 miles deep, we should be able to dilute effectively all the garbage thrown into it. But actually, about 95 percent of the total air mass is within only 12 miles of the surface. This layer, called the *troposphere,* is where we have our weather and air pollution problems.

The science of meteorology has great bearing on air pollution. An air pollution problem involves three parts: the source, the movement of the pollutant, and the recipient (Figure 18–1). Whereas Chapter 17 covers sources and effects, this chapter covers the transport mechanism—how the pollutants travel through the atmosphere.

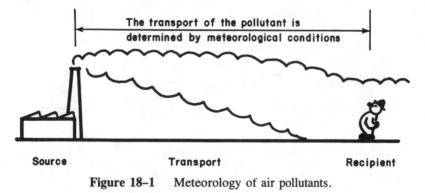

The transport of the pollutant is determined by meteorological conditions

| Source | Transport | Recipient |

Figure 18-1   Meteorology of air pollutants.

BASIC METEOROLOGY

Weather changes in the form of fronts, which can be warm or cold. Warm fronts usually are associated with steady rain and drizzle; cold fronts bring heavy local rain.

Another way of picturing weather is in terms of barometric pressure. Low pressure systems are associated with both hot and cold fronts. The air movement around low pressure systems is counterclockwise (in the northern hemisphere) and vertical winds are upward where condensation and precipitation take place. High pressure systems bring sunny and calm weather with the wind spiraling clockwise and downward. There are no fronts associated with high pressure systems, which represent stable atmospheric conditions. The low and high pressure systems, commonly called *cyclones* and *anticyclones* respectively, are illustrated in Figure 18–2. Because of the high stability in anticyclones, these are usually precursors to air pollution episodes.

A primary objective of air quality management is to attain low air pollutant concentrations in community air. This can be accomplished of course by reducing the amount of pollutants emitted. It is also possible to reduce concentrations by attaining adequate dispersal and dilution in the atmosphere. Such dispersal is controlled by atmospheric conditions and can be vertical as well as horizontal. The principles governing both vertical and horizontal movement of air are therefore important in air quality management.

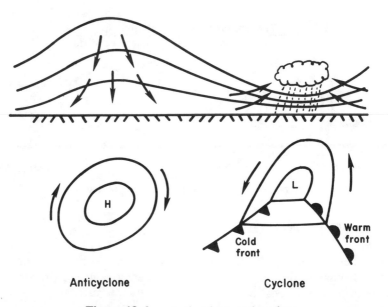

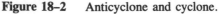

Anticyclone   Cyclone

Figure 18–2   Anticyclone and cyclone.

HORIZONTAL DISPERSION OF POLLUTANTS

The earth acts like a wave converter, accepting the sun's light energy (high-frequency waves) and converting it to heat energy (low-frequency waves), which is then radiated back to space. The heat transfer from the earth to space is by radiation, conduction, and convection. Radiation involves the transfer of heat by energy waves, a minor effect on the atmosphere; *conduction* is the transfer of heat by physical contact; *convection* is the process of heating by the movement of air masses.

If the earth did not rotate, the air near the equator would be heated; then it would rise and move toward the poles where it would sink (Figure 18–3). But the earth does rotate and thus always presents new areas for the sun to shine on and to warm. Accordingly, a pattern of winds is set up around the world, some seasonal (for instance, hurricanes) and some permanent. Local conditions and cloud cover further complicate the picture.

For example, land masses heat and cool faster than water; thus shoreline winds blow out to sea during the night and inland during the day. Valley and slope winds are caused by the cooling of air on mountain slopes. In cities, brick and stone absorb and hold heat, creating a *heat island* about the city during the night (Figure 18–4). This effect sets up a self-contained circulation pattern called a *haze hood* from which the pollutants cannot escape.

The horizontal motion of winds is measured as wind velocity. These wind velocity data are plotted as a *wind rose,* a graphic picture of the direction and velocity *from which the wind came.* The wind rose in Figure 18–5 shows that the prevailing winds were from the southwest.

Air pollution engineers often use a variation of the wind rose, called a *pollution rose,* to determine the source of a pollutant. Instead of plotting all winds on a radial graph, only those days during which the concentration of a pollutant is above a certain minimum are used. Figure 18–6 is an actual plotting of such a pollution rose. Only winds carrying SO_2 levels greater than 250 $\mu g/m^3$ were plotted. Note how the fingers of the rose point to Plant 3. Pollution roses can be plotted for other pollutants as well and are useful for pinpointing sources of atmospheric contamination.

VERTICAL DISPERSION OF POLLUTANTS

As a parcel of air rises in the earth's atmosphere, it experiences lower and lower pressure from surrounding air molecules, and thus it expands.

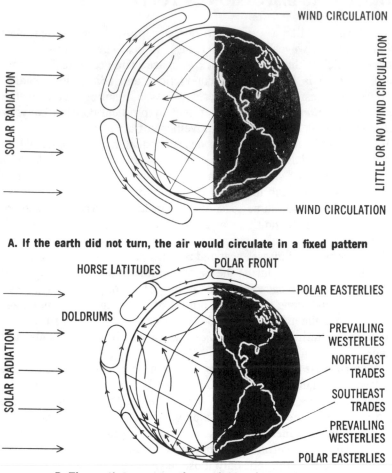

A. If the earth did not turn, the air would circulate in a fixed pattern

B. The earth turns, creating variable wind patterns

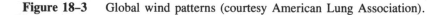

Figure 18-3 Global wind patterns (courtesy American Lung Association).

This expansion lowers the temperature of the air parcel. The rate at which dry air cools as it rises is called the *dry adiabatic lapse rate* and is independent of the ambient air temperature. The term *adiabatic* means that there is no heat exchange between the parcel of air under consideration and the surrounding air. The dry adiabatic lapse rate may be calculated directly from the first law of thermodynamics as

$$-1\,°C/100\ m = \frac{-5.4\,°F}{1000\ ft}$$

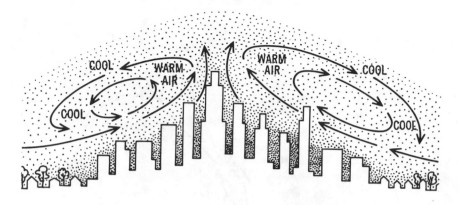

Figure 18-4   Heat island formed over a city.

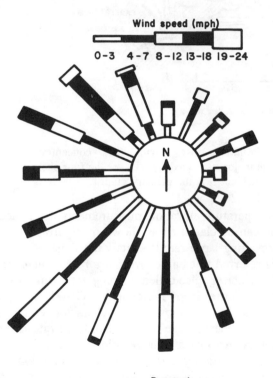

Figure 18-5   Typical wind rose.

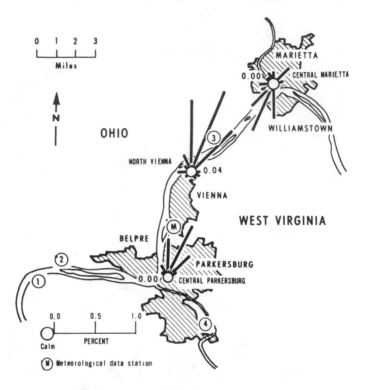

Figure 18–6   Pollution roses, with SO_2 concentrations greater than 250 μg/m^3. The major suspected sources are the four chemical plants, but the data indicate that Plant 3 is the primary culprit.

The actual temperature-elevation measurements are called *prevailing lapse rates* and can be classified as shown in Figure 18–7.

When the prevailing lapse rate is exactly the same as the dry adiabatic lapse rate, it is referred to as a *neutral lapse rate* or *neutral stability*. A *superadiabatic lapse rate*, also called *a strong lapse rate*, occurs when the atmospheric temperature drops more than 1 °C/100 m. A *subadiabatic lapse rate*, also called a *weak lapse rate*, is characterized by a drop of less than 1 °C/100 m. A special case of the weak lapse rate is the *inversion*, a condition that has warmer air above colder air.

During a superadiabatic lapse the atmospheric conditions are unstable; a subadiabatic and especially an inversion characterizes a stable atmosphere. This can be demonstrated by depicting a parcel of air at 500 m (see Figure 18–8). If the temperature at 500 m is 20 °C, during a superadiabatic condition the temperature at ground level might be 30° and at 1000 m it might be 10° (Note: a change of more than 1 °C/100 m).

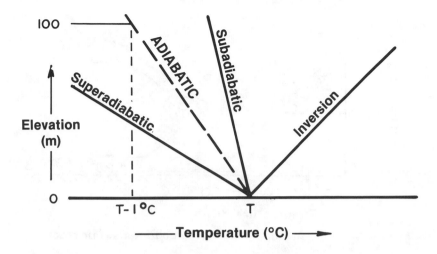

Figure 18–7   Prevailing lapse rates, and the dry adiabatic lapse rate.

If the parcel of air at 500 m is moved upward to 1000 m, what would its temperature be? Remember that, assuming adiabatic condition, the parcel would cool 1 °C/100 m. The temperature of the parcel at 1000 m is thus 5° less than 20° or 15 °C.

The prevailing temperature, however, is 10 °C, and the air parcel finds itself surrounded by cooler air. Will it rise or fall? Obviously it will rise since warm air rises. Once the parcel of air under superadiabatic conditions is displaced upward, it keeps right on going.

Similarly, if a parcel is displaced downward, say to ground level, the air parcel is 20° + (500 m × [1 °/100 m]) = 25 °C. It finds the air around it a warm 30°, and thus the cooler air parcel would just as soon keep going down if it could. Superadiabatic conditions are thus unstable, characterized by a great deal of vertical air movement and turbulence.

The subadiabatic condition shown in Figure 18–8 is by contrast a very stable system. Consider again a parcel of air at 500 m and at 20 °C. A typical subadiabatic system has ground level temperature of 21 °C and 19 °C at 1000 m. If the parcel is displaced to 1000 m, it will cool by 5–15°. But finding the air around it a warmer 19°, it will fall right back to its point of origin. Similarly, if the air parcel were brought to ground level, it would be at 25°, and finding itself surrounded by 21° air, it would rise back to 500 m. Thus the subadiabatic system would tend to dampen out vertical movement and is characterized by a very limited vertical mixing.

An inversion is an extreme subadiabatic condition, and thus the vertical air movement within an inversion is almost nil.

In Los Angeles, the inversion is called a *subsidence inversion* and is

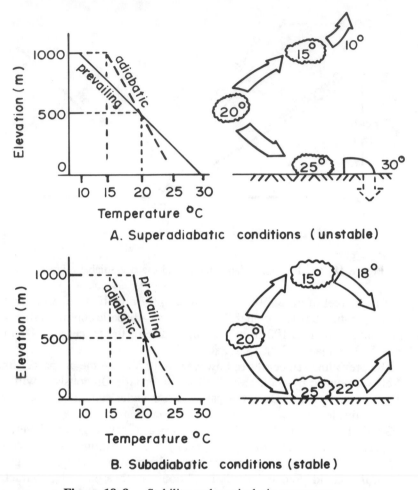

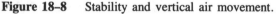

Figure 18-8   Stability and vertical air movement.

caused by a large warm air mass subsiding over the city. A more common type of inversion is the *radiation inversion,* caused by the radiation of heat to the atmosphere from the earth. During the night as the earth cools, the air close to the ground loses heat, thus causing an inversion (Figure 18-9). The pollution emitted during the night is caught under this lid and does not escape until the earth warms sufficiently to break the inversion. Cities in inland mountain valleys like Denver and Santa Fe experience frequent radiation inversions.

Atmospheric stability can often be recognized by the shapes of plumes emitted from smokestacks (Figure 18-10). One potentially serious condition is called *fumigation*—where the pollutants are caught under an inver-

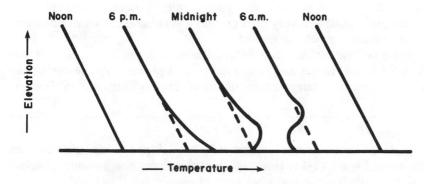

Figure 18–9   Typical prevailing lapse rates during a sunny day and clear night.

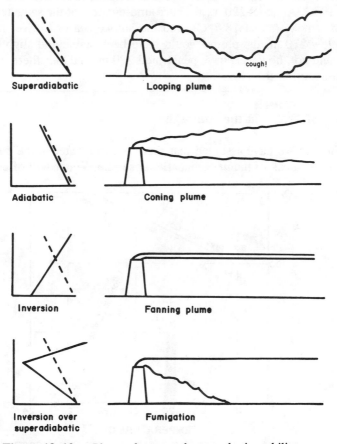

Figure 18–10   Plume shapes and atmospheric stability.

sion and are mixed due to a strong lapse rate. A looping plume also can be dangerous due to a very high ground level concentration of pollutants as the plume touches the ground.

If we assume perfect adiabatic conditions in a plume, we can estimate how far it will rise (or sink) and what type of plume it will be during any given atmospheric temperature condition. This is illustrated in Example 18.1.

Example 18.1

A stack100 m tall emits a plume at 20 °C. The prevailing lapse rates are shown in Figure 18–11. How high will the plume rise (assuming perfect adiabatic conditions), and what type of plume will it be?

Note that the prevailing lapse rate is subadiabatic to 200 m and an inversion exists above 200 m. The smoke at 20 °C finds itself surrounded by colder (18.5 °C) air and thus rises. As it rises, it cools so that at 200 m it is 19 °C. At about 220 m, the surrounding air is at the same temperature as the smoke (about 18.7 °C), and the smoke ceases to rise.

Below 220 m, the plume would have been stable and slightly coning. It would not, however, have penetrated 220 m, and thus there would have been a cap on the plume.

Effect of Water in the Atmosphere

Thus far we have assumed that there was no water in the atmosphere, hence we speak of the *dry* adiabatic lapse rate. Water will of course con-

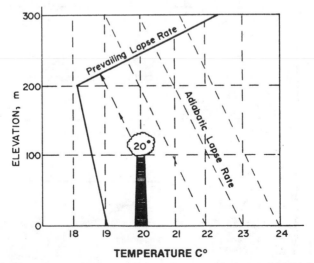

Figure 18–11   Atmospheric conditions for an emission at 20 °C, for Example 18.1

dense and evaporate, and in so doing emit and absorb heat, making the calculations of stability quite a bit more complicated. In general as a parcel of air rises, the water vapor in that parcel will condense, releasing heat. The wet adiabatic lapse rate will thus be less negative than the dry adiabatic lapse rate. The wet adiabatic lapse rate has been observed to vary from $-0.6\,°C/100$ m to $-0.3\,°C/100$ m.

Water also affects air quality in other ways. Fogs are formed when moist air cools, and the moisture condenses. Aerosols provide the condensation nuclei, and fogs thus tend to occur more readily in urban areas.

In addition to inversions, serious air pollution episodes are almost always accompanied by fogs. These tiny droplets of water are detrimental in two ways. In the first place, fog makes it possible to convert SO_3 to H_2SO_4. Secondly, fog sits in valleys and prevents the sun from warming the valley floor and breaking inversions, thus often prolonging air pollution episodes.

ATMOSPHERIC DISPERSION

When a plume containing pollutants comes out of the stack into the air, the pollutants are dispersed as the plume spreads out through the atmosphere. The concentration of air pollutants at ground level is determined by this dispersion as much as by the rate of emission from the stack.

The plume from a stack moves mainly downwind, and the greatest concentration of pollutant molecules is always along the plume centerline. There is also, however, diffusion of molecules across (perpendicular to) the wind direction, and the plume thus spreads out laterally as it moves downwind. This phenomenon is shown in Figure 18–12.

A reasonably good model of air pollution dispersion is given if we assume that the vertical and horizontal cross sections of the plume have the traditional Gaussian shape. Complete analysis of pollution dispersion is beyond the scope of this text. Two results of this analysis, however, are given below.

For a source of pollutant emission at ground level the ambient pollutant concentration directly downwind from the source is given by

$$\chi(x, 0, 0) = Q/\pi u \sigma_y \sigma_z$$

where $\chi(x, 0, 0)$ = ambient concentration at ground level, downwind along the plume centerline at distance x

         $\bar{u}$ = average wind speed

       σ_y, σ_z = standard deviation of the plume in the lateral and vertical directions

         Q = rate of pollutant emission

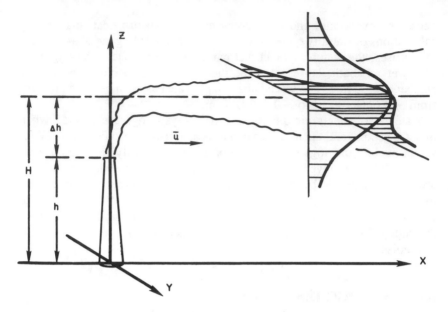

Figure 18–12   Gaussian dispersion model.

If Q is measured in g/sec, the wind speed, in m/sec, and the standard deviations in meters, the units of the concentration will be g/m^3. The maximum concentration downwind at ground level along the plume centerline *for a plume emitted either at ground level or from a stack* is given by:

$$\chi_{max} = 0.1171Q/\bar{u}\sigma_y\sigma_z$$

The standard deviations σ_y and σ_z are measures of the plume spread in the lateral and vertical directions. They depend both on atmospheric stability and on distance from the source. Stability is classified in categories A–F, called *stability classes*. Table 18–1 provides a key to the stability classes and the wind speeds that are consistent with each. In terms of lapse rates, classes A and B are associated with superadiabatic, unstable meteorological conditions; class C, with near neutral stability; class D, with subadiabatic, slightly stable conditions; and classes E and F, with weak and strong inversions respectively. We should note that urban, populated regions rarely achieve greater stability than class D because of the heat island effect. Stability classes E and F are generally found in rural or unpopulated areas.

The variability of σ_y and σ_z with distance for the various atmospheric stability categories is shown in Figure 18–13.

Table 18–1. Atmospheric Stability Key for Figure 18–13

Surface Wind Speed (at 10m) (m/sec)	Day[a] Incoming Solar Radiation (Sunshine)			Night[a]	
	Strong	Moderate	Slight	Thinly Overcast or 4/8 Low Cloud	3/8 Cloud
<2	A	A–B	B	—	—
2–3	A–B	B	C	E	F
3–5	B	B–C	C	D*	E
5–6	C	C–D	D	D	D
>6	C	D	D	D	D

[a] The neutral category, D, should be assumed for overcast conditions during day or night.

Example 18.2

An oil pipe leak results in emission of 100 g/hr of H_2S. When the wind speed is 3.0 m/sec and under stability class C, what will be the concentration of H_2S 1.5 km downwind from the leak?

From Figure 18–13 at x = 1500 meters,

$$\sigma_y = 160 \text{ m and } \sigma_z = 60 \text{ m}$$

$$Q = 100 \text{ g/hr} = 2.8 \text{ g/sec}$$

$$C(1500,0,0) = (2.8 \text{ g/s})/\pi(3.0\text{m/s})(160\text{m})(60\text{m})$$

$$= 1.7 \times 10^{-5} \text{ g/m}^3 \text{ or } 17 \ \mu\text{g/m}^3$$

CLEANSING THE ATMOSPHERE

Since we obviously have not yet all suffocated even though prodigious amounts of pollutants have been thrown into the air over the millions of years, there must exist a series of processes by which air is cleansed. These include the effect of gravity, contact with the earth's surface, and removal by precipitation.

Particulates, if of sufficient size, are removed by simple settling out under the influence of gravity. Unfortunately, the settling velocities of common particulates are very small. For example, a 1-μm particle will have a settling velocity of about 1 cm/sec in ideal quiescent conditions. Practically, due to turbulence in the atmosphere, particles smaller than 20 μm will seldom settle out by gravity. Gases are removed by gravity

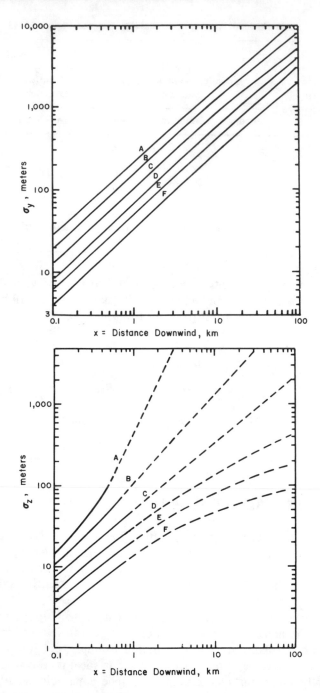

Figure 18–13   Dispersion coefficients.

only if they are adsorbed onto particulates. Sulfur dioxide, for example, is readily adsorbed and is thus partially removed by gravity.

Particulate matter is not dispersed in the air in precisely the same way as gaseous pollutants are.

Particles larger than a millimeter in diameter fall out of the air too rapidly to be dispersed over any great distance. The particles of concern as air pollutants, which are small enough to stay in the air for appreciable periods of time but large enough to be dispersed differently from gases, are generally between 1 μm and 100 μm in diameter.

Many of the gases are adsorbed by the earth's surface, including stone, vegetation and other materials. Some gases such as SO_2 are readily dissolved in surface waters.

The third major removal mechanism is by precipitation. Two types of removal occur, the first being an "in-cloud" process called *rainout* where submicron particles become nuclei for the formation of rain droplets that will grow and eventually fall as precipitation. The second mechanism is called *washout* and is a "below-cloud" process where the rain falls through the air pollutants, the pollutants are impinged or dissolved in the droplets, and then carried to earth.

The relative importance of these removal mechanisms was illustrated by a study of SO_2 emissions in Great Britain, where the surface sink accounted for 60 percent of the SO_2, 15 percent was removed by precipitation and 25 percent left Great Britain (heading you-know-where).

CONCLUSION

Air pollution episodes are the results of high emissions and a combination of meteorological factors. Some of these factors are:

1. little horizontal wind movement
2. stable atmospheric conditions resulting in very limited vertical air movement
3. fog, which promotes the formation of secondary pollutants and hinders the sun from warming the ground and breaking inversions
4. high pressure areas resulting in downward vertical air movement and absence of rain for washing the atmosphere

It would seem reasonable therefore that episodes can be predicted on the basis of meteorological data, provided the potential exists. The EPA in cooperation with the Weather Bureau has indeed established procedures for evaluating meteorological data so as to provide early warning for impending episode conditions and has developed emergency plans (including shutting down industries) should the conditions warrant.

PROBLEMS

18.1 Given the following temperature soundings:

Elevation (m)	Temperature (°C)
0	20
50	15
100	10
150	15
200	20
250	15
300	20

what type of plume would you expect if the exit temperature of the plume were 15°C and the smoke stack were
  a.   50 m tall?
  b.   150 m tall?
  c.   250 m tall?

18.2 Consider a prevailing lapse rate that has these temperatures: ground =21°C, 500 m = 20°C, 600 m = 19°C, and 1000 m = 20°C. If we released a parcel of air at 500 m and at 20°C, would it tend to sink, rise, or remain where it was? If a stack is 500 m tall, what type of plume would you expect to see?

18.3 Draw a map with x and y coordinates (x = horizontal, y = vertical) and place on the map the following:
  Industrial Plant "A" at $x = 3$, $y = 3$
  Industrial Plant "B" at $x = 8$, $y = 1$
  Industrial Plant "C" at $x = 8$, $y = 8$
  Air Sampling Station at $x = 5$, $y = 5$

The data at the air sampling station are:

Day	Wind Direction	Particulates ($\mu g/m^3$)	SO_2 $\mu g/m^3$
1	N	80	80
2	NE	120	20
3	NW	30	30
4	N	90	40
5	NE	130	20
6	SW	20	180
7	S	30	100

8	SW	40	200
9	E	100	60
10	W	10	100

Draw pollution roses to show which plant is guilty of the air pollution.

18.4 If your job were to continuously analyze meteorological data and watch for conditions that might lead to the occurrence of an "episode," what specific conditions would you be looking for? (That is, what are the meteorological criteria for the formation of an episode?)

18.5 A power plant burns 1000 tons of coal/day, 2 percent of which is sulfur. This is emitted from the 100-m stack. For a wind speed of 10m/sec, calculate: (a) the maximum ground level concentration of SO_2 10 km downwind from the plant, (b) the maximum ground level concentration and the point at which this occurs for stability categories A, C, and F.

18.6 Given a wind velocity of 2 m/sec, calculate the maximum expected SO_2 concentration at ground level in a town 10 km downwind from a power plant that burns 1000 metric tons of coal per day (sulfur content of 1 percent). The stack is 100 m high.

18.7 Consider the following atmospheric temperature soundings:

Elevation (ft)	Temperature (°F)
0	70.0
200	68.0
400	66.0
600	72.0
800	70.0
1000	68.0

a. Indicate below the type of lapse rate involved
     0 to 400 ft   _____
     400 to 600 ft   _____
     600 to 1000ft   _____
b. Indicate below the plume type if the stack were
     300 ft tall   _____
     500 ft tall   _____
     700 ft tall   _____
c. What would be the ground level concentration of the pollutant (negligible, moderate, high) if the stack were
     300 ft tall   _____
     500 ft tall   _____
     700 ft tall   _____

18.8 The odor threshold of H_2S is about 0.7 $\mu g/m^3$. If an industry emits 0.08 g/sec of H_2S out of a 40-m stack during an overcast night with a wind speed of 3 m/sec, estimate the area (in terms of x and y coordinates) where H_2S would be detected. (Use a computer.)

18.9 The concentration of H_2S at a location 200 m downward from an abandoned oil well is 3.2 $\mu g/m^3$. What is the emission rate, on a partially overcast afternoon if the wind speed is 2.5 m/sec?

LIST OF SYMBOLS

H = effective height of the stack, m

h = stack elevation, m

Q = emission rate, kg/sec

T = temperature

$\bar{u}$ = average wind speed, m/sec

σ_y = standard deviation, y direction, m

σ_z = standard deviation, z direction, m

χ = concentration of pollutant, kg/m^3

Appendix:
Atmospheric Dispersion

Dispersion is the process of spreading out the emission over a large area and thus reducing the concentration of the specific pollutants. The plume spread or dispersion is in two dimensions: horizontally and vertically. We assume that the greatest concentration in the pollutants is in the plume centerline, that is, in the direction of the prevailing wind. The further we get from the centerline, the lower the concentration. If we assume that the spread of a plume in both directions is approximated by a Gaussian probability curve, we can calculate the concentration of a pollutant at any distance x downwind from the source by

$$\chi_{(x,y,z)} = \frac{Q}{2\pi \bar{u} \sigma_y \sigma_z} \exp(-\tfrac{1}{2}[(y/\sigma_y)^2 + (z/\sigma_z)^2])$$

where

χ = concentration at some point in the x, y, z coordinate space, kg/m^3

Q = emissions, kg/sec

$\bar{u}$ = average wind speed, m/sec

σ_y and σ_z = standard deviation of the dispersion in the y and z directions

The coordinates are shown in Figure 18–12. Note that z is in the vertical direction, y is horizontal crosswind, and x is downwind.

The standard deviations are measures of how much the plume spreads. If y and z are large, the spread is great, and the concentration is of course low. The opposite is true if the spread is small.

The dispersion is dependent on both atmospheric stability and the distance from the source. Figure 18–13 is one approximation for the dispersion coefficients.

Atmospheric stability is denoted in Figure 18–13 by letters ranging from A to F. Table 18–1 is a key for selecting the proper stability condition.

Suppose a stack has an effective height (stack height plus plume rise) of H meters. The elevation of the plume centerline is $z = H$, and the diffusion equation is thus

$$\chi_{(x,\,y,\,z)} = \frac{Q}{2\pi\bar{u}\sigma_y\sigma_z}\,\exp(-\tfrac{1}{2}[(y/\sigma_y)^2 + ((z - H)/\sigma_z)^2])$$

Note that at the plume centerline elevation $z = H$ and the last exponential term drops out, yielding the equation

$$\chi_{(x,\,y,\,z)} = \frac{Q}{2\pi\bar{u}\sigma_y\sigma_z}\,\exp(-\tfrac{1}{2}[(y/\sigma_y)^2])$$

These equations hold as long as the ground does not influence the diffusion. This is usually not a good assumption since the ground is not a 100 percent efficient sink for the pollutants, and the levels must be higher at ground level due to inability of the plume to disperse into the ground. This effect can be taken into account if we think of an imaginary mirror image source at elevation $z - H$, as shown in Figure 18A–1. Taking this into account, we can write

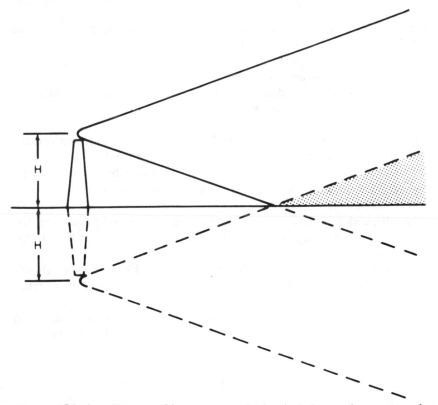

Figure 18A–1   Source and image source. In the shaded area, the concentration is doubled due to the image source.

$$\chi_{(x, y, z)} = \frac{Q}{2\pi \bar{u} \sigma_y \sigma_z} (\exp(-\frac{1}{2}[(y/\sigma_y)^2]))$$

$$\times (\exp(-\frac{1}{2}[(z - H)^2/\sigma_z^2]) + \exp(-\frac{1}{2}[(z + H)^2/\sigma_z^2]))$$

Example 18A.1

Given a sunny summer afternoon with average wind, $\bar{u} = 4$ m/sec, emission $Q = 0.01$ kg/sec, and the effective stack height 20 m, find the ground level concentration at 200 meters from the stack.

Using the above equation and from Figure 18–12 finding that at 200 meters, $\sigma_y = 36$ m and $\sigma_z = 20$ m for an unstable superadiabatic strong solar radiation (Table 18–1), the atmospheric conditions are (Type B), and noting that maximum concentrations occur on the plume centerline, at $y = 0$

$$\chi = \frac{Q}{2\pi \bar{u} \sigma_y \sigma_z} (\exp(-\frac{1}{2}[(y/\sigma_y)^2]))$$

$$\times (\exp(-\frac{1}{2}[(z - H)^2/\sigma_z^2]) + \exp(-\frac{1}{2}[(z + H)^2/\sigma_z^2]))$$

$$\chi = \frac{0.01}{2(3.14)(4)(36)(20)} (\exp -\frac{1}{2}[(0/36)^2]))$$

$$\times (\exp(-\frac{1}{2}[(0 - 20)^2/20^2]) + \exp(-\frac{1}{2}[(0 + 20)^2/20^2]))$$

$$\chi = (5.53 \times 10^{-7})(1)(e^{-\frac{1}{2}} + e^{-\frac{1}{2}})$$

$$\chi = (5.53 \times 10^{-7})(0.6 + 0.6) = 6.64 \times 10^{-7} \text{ kg/m}^3$$

$$\chi = 664 \mu g/m^3$$

Finally, it should be pointed out that the accuracy of this plume rise and dispersion analysis is very poor. Air pollution modelers are usually pleased to find their models predicting concentrations to within an order of magnitude! We thus should refrain from placing undue validity on these models.

Chapter 19

Measurement of Air Quality

It is previously observed that air quality measurement, like water quality measurement, is complicated by the lack of knowledge as to what is "clean" and by the difficulty in defining "quality." "Pure air" is ordinarily defined as containing only the naturally occurring gases, but "pure air" does not exist in nature. Pollen, dust, fog, and so forth are all contaminants although they occur without any assistance from industrialized man. No attempt is usually made to differentiate between natural and people-made pollutants and air quality measurements are designed to measure all types of contaminants. Air quality measurements also involve very small concentrations. Such measurements are inherently difficult.

The measurements of air quality generally fall into three classes:

1. *Measurement of emissions.* This is called *stack sampling* when a stationary source is analyzed. A hole is punched into the stack and samples drawn out for on-the-spot analyses. Sampling moving sources is more difficult.
2. *Meteorological measurements.* The measurement of meteorological factors is necessary if we are to know how and why the pollutants travel from the source to the recipient. Some of these are discussed in Chapter 18.
3. *Ambient air quality.* The quality of ambient air is of course of major concern. Almost all evidence of health effects is based on these measurements. Monitoring air quality can also provide data for recognizing episodes and thus can provide some warning of impending health problems.

MEASUREMENT OF PARTICULATES

First-generation devices for measuring particulates involved the determination of how much dust settles to the earth. The measurement of particulates by dustfall jars is subject to many problems (not the least of which is uncooperative pigeons). The second-generation particulate sampling device, the *high-volume sampler* (or *hi-vol*), the workhorse of particulate sampling (Figure 19–1), operates much like a vacuum cleaner by simply forcing over 2000 cubic meters (70,000 ft^3) of air through the filter in 24 hours. The analysis is gravimetric. The filter is weighed before and after, and the difference is the weight of the particulates collected.

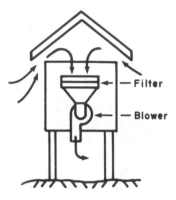

Figure 19–1   Hi-vol sampler.

The air flow is measured by a small flow meter, usually calibrated in cubic feet of air per minute. Because the filter gets dirty during the 24 hours of operation, less air goes through the filter during the latter part of the test than at the beginning and the air flow must therefore be measured at both the start and end of the test period and the values averaged.

Example 19.1

A clean filter is found to weigh 10.00 grams. After 24 hours in a hi-vol, the filter plus dust weighs 10.10 grams. The air flow at the start and end of the test was 60 and 40 ft^3/minute respectively. What is the particulate concentration?

$$\text{Weight of the particulates (dust)} = (10.10 - 10.00)\text{g} \times 10^6 \ \mu\text{g/g}$$

$$= 0.1 \times 10^6 \ \mu\text{g/g}$$

$$\text{Average air flow} = \frac{60 + 40}{2} = 50 \ \text{ft}^3/\text{min}$$

$$\text{Total air through the filter} = 50 \ \text{ft}^3\text{min} \times 60 \ \text{min/hr} \times 24 \ \text{hr/day} \times 1 \ \text{day}$$

$$= 72{,}000 \ \text{ft}^3$$

$$= 72{,}000 \ \text{ft}^3 \times (28.3 \times 10^{-3}) \ \text{m}^3/\text{ft}^3$$

$$= 2038 \ \text{m}^3$$

$$\text{Total suspended particulates} = \frac{0.1 \times 10^6 \mu\text{g}}{2038 \ \text{m}^3}$$

$$= \ 49 \ \mu\text{g/m}^3$$

The particulate concentration thus measured is often referred to as *total suspended particulates* (*TSP*) to differentiate it from other measurements of particulates.

Another widely used measure of particulates in the environmental health area is of *respirable particulates,* or those particulates that would be respired into lungs. These are generally defined as being less than 0.3 μ in size, and the measurements are done with stacked filters. The first filter removes only particulates $>0.3 \ \mu$, and the second filter, having smaller spaces, removes the smaller respirable particulates.

Interestingly, no third-generation particulate measuring devices have yet been developed and accepted. The difficulty, of course, is that the measurement must be gravimetric, and it is difficult to construct a device that continuously weighs minute quantities of dust.

It is often necessary to measure the decrease in visibility as a part of air quality assessment. Fine particles interfere with visibility by scattering light. This scattered light can be measured with a *nephelometer,* an instrument that measures emitted or scattered light at 90° to incident light. The air nephelometer can be calibrated directly in units of percent visibility decrease.

MEASUREMENT OF GASES

Sulfur dioxide is measured by impregnating filter papers with chemicals that react with SO_2 and change color. For example, lead peroxide reacts as

$$PbO_2 + SO_2 \rightarrow PbSO_4$$

forming a dark lead sulfate. The extent of this reaction is estimated by the dark areas on the filter paper.

Many techniques involve the use of a *bubbler,* shown in Figure 19–2. The gas is literally bubbled through the liquid, which either reacts chemically with the gas of interest or into which the gas is dissolved. Wet chemical techniques are then used to measure the concentration of the gas.

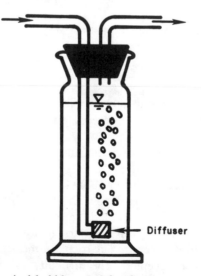

Figure 19–2   A typical bubbler used for the measurement of gaseous air pollutants.

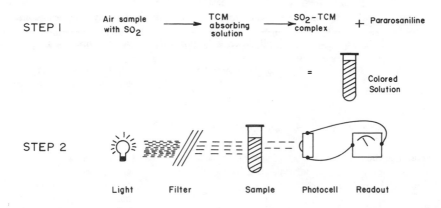

Figure 19-3   Schematic of the pararosaniline method for measuring SO_2.

One of the better methods of measuring SO_2 is the colorimetric pararosaniline method, in which SO_2 is bubbled into a liquid containing tetrachloromercurate (TCM). The SO_2 and TCM combine to form a stable complex. Pararosaniline is then added to this complex with which it forms a colored solution. The amount of color is proportional to the SO_2 in the solution, and the color is measured with a spectrophotometer at a wavelength of 560 mm. (See Chapter 3 for ammonia measurement—another example of a colorimetric technique.) Figure 19-3 graphically illustrates the pararosaniline method for SO_2 measurement.

Most bubblers are not 100 percent efficient for not all of the gas bubbled in will be absorbed by the liquid and some will escape. Obviously, this creates problems since it is usually necessary to have quantitative determinations. Most bubblers are thus tested to establish their efficiencies, and these values are used as contrasts by which the measured values are multiplied to obtain the actual concentrations. Gas chromatographic techniques are being developed for SO_2 assay.

Hundreds of different techniques have been used in third-generation gaseous measurement. One widely used device is the *nondispersive infrared analyzer,* used for carbon monoxide measurement. This technique relies on the fact that CO absorbs infrared radiation of a specific frequency. Nitrogen oxides and photochemical oxidants are measured using a technique called chemiluminescence.

REFERENCE METHODS

Because of the great number of methods available for measuring air pollutants, the U.S. Environmental Protection Agency has chosen a

Table 19–1.   Standard EPA Reference Methods for Air Quality Measurements

Pollutant	Reference Method	Comments
Particulates (TSP)	High-vol sampler	Note that is a second-generation device, giving 24-hr readings.
Sulfur dioxide (SO_2)	Pararosaniline	
Carbon monoxide (CO)	Nondispersive infrared spectrometry	
Nitrogen dioxide (NO_2)	Chemiluminescence	This method was adopted after the original reference method was judged unreliable (and years of NO_2 monitoring data became worthless).
Photochemical oxidants (O_3)	Chemiluminescence	This method measures oxidants as O_3.
Hydrocarbons (nonmethane)	Flame ionization	Methane is not measured because it is a nonreactive hydrocarbon, naturally occurring.

series of reference methods (Table 19–1). These are not necessarily absolutely accurate, but they are judged the best available, and results from other methods are to be compared to these. The existence of reference methods is especially important when compliance with air quality standards is in question.

GRAB SAMPLES

Often it is necessary to obtain a sample of a gas for future analysis in the laboratory. Whereas in water pollution work the collection of samples does not present a serious problem, in air quality measurements it is a real challenge.

Plastic and aluminum bags have been used extensively. Usually the gas is pumped in and allowed to escape through a hole. By thus displacing two or three volumes of the bag, contamination problems are avoided.

STACK SAMPLING

Stack sampling is an art worthy of individual attention. As the name implies, this involves the sampling of gas from a smokestack and is neces-

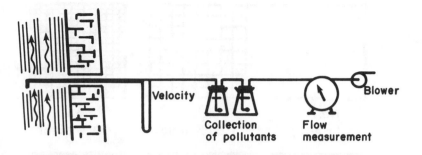

Figure 19-4 A stack sampling train.

sary for evaluating compliance with emission standards and determining efficiencies of air pollution control equipment.

The most serious problem with stack sampling is the risk of obtaining an unrepresentative sample. Accordingly, a thorough survey is usually made of the flow, temperature, and pollutant concentration in a stack. It is also usually desirable to obtain samples from a series of locations within a stack to be assured of accurate measurements. A train of instruments (Figure 19-4) is often utilized for stack sampling so that a number of measurements can be determined at each positioning of the intake nozzle.

SMOKE AND OPACITY

Air pollution has historically been associated with smoke—the darker the smoke, the more the pollution. We now of course know that this is not necessarily true, but many regulations (for example, those for municipal incinerators) are still written on the basis of smoke density. The opacity of a smoke plume is still the only method for enforcing air quality standards that can be used without the emitter's knowledge. Most citations for air pollution violations are written for opacity violations.

The density of black or gray smoke is measured on the Ringelmann scale. One end of the scale, Ringelmann 0, is complete transparency (no visible plume); the other end, Ringelmann 5, is a completely opaque black plume. A typical opacity standard for a stationary source is Ringelmann 1 (20 percent opacity) with allowances for very short periods of Ringelmann 2 (40 percent opacity). Ringelmann 1 is a barely visible plume. The Ringelmann test was at one time conducted by comparing the blackness of a card (such as those shown in Figure 19-5) to the blackness of the observed plume. Modern practice involves the training of enforcement agents (in a "smoke school") to recognize Ringelmann opacities

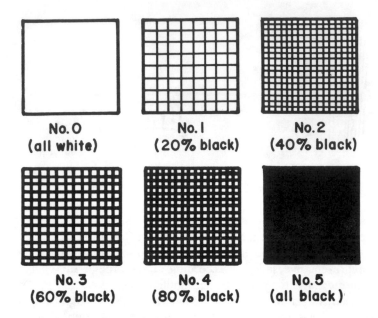

No. 0
(all white)

No. I
(20% black)

No. 2
(40% black)

No. 3
(60% black)

No. 4
(80% black)

No. 5
(all black)

Figure 19–5   Ringelmann scale for measuring the opacity of smoke.

by repeated observation of smoke of predetermined opacity. The opacity of white smoke is reported as "percent opacity" rather than by Ringelmann number. Opacity may also be measured continuously by installing a photometer in the stack breach and calibrating the emitted smoke by the Ringelmann or opacity scale.

CONCLUSION

As with water pollution, the analytical tests of air quality can be only as good as the samples or sampling techniques used. In addition, the prevailing analytical techniques leave a great deal to be desired in both precision and accuracy. It is important to remember, therefore, that most measurements of environmental quality, especially of air quality, are at best reasonable estimates and should not be believed to the fourth decimal point.

PROBLEMS

19.1 A hi-vol clean filter weighs 20.0 g, and the dirty filter weighs 20.5 g. The initial and final air flows were 70 and 50 ft³/min. What volume of air went through the filter in 24 hours? What was the level of particulates in the atmosphere? How does this compare to the National Air Quality Standards? (See Chapter 21.)

19.2 A high-vol sampler draws air in at an average rate of 70 ft^3/min. If the particulate reading is 200 μg/m^3, what was the weight of the dust on the filter?

Chapter 20

Air Pollution Control

It is not difficult to see how air has become the ubiquitous wastebasket. For ages people have been dumping wastes into the atmosphere, and these pollutants have "disappeared" with the wind. It is difficult, therefore, to suddenly force people to limit emissions and almost impossible to get them to pay for it. Nevertheless, our industrialized society must now prevent the emission of what would a few years ago have been wasted into the atmosphere.

The control of emissions can be realized in a number of ways. Five separate possibilities for control are pictured in Figure 20-1. Dispersion is covered in Chapter 18, and the four remaining control methods are discussed individually in this chapter.

SOURCE CORRECTION

Often the easiest solution to an air pollution problem is to stop or change the guilty process. Once the decision has been made that a product or process is necessary, the engineer must consider the possibility of controlling emissions by changing the process. For example, if automobiles are blamed for high lead levels in urban air, the most reasonable solution is elimination of the lead from gasoline. Similarly, removal of sulfur from coal and oil is possible before the fuel is burned. In these cases, the source has been corrected and the problem solved.

In addition to a change of raw material, a modification of the process might also be used to achieve a desired result. For example, municipal refuse incinerators have been known to stink. The odors can often be readily controlled if the incinerators are operated at a high enough temperature to completely oxidize the organics that cause the odor. Pro-

299

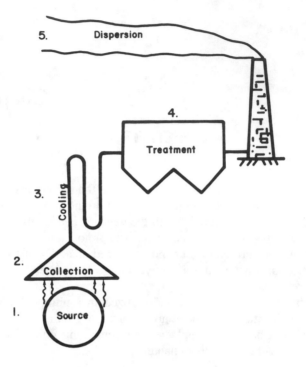

Figure 20–1   Points of possible air pollution control.

vided this higher-temperature operation is possible, it is a very reasonable process change to obtain the desired air pollution control.

Strictly speaking, such measures as process change, raw material conversion, or equipment modification to meet emission standards are known as *controls*. In contrast, *abatement* is the term used for all devices and methods for decreasing the quantity of pollutant reaching the atmosphere once it has already been emitted from the source. For the sake of simplicity, however, we refer to all of the procedures as controls.

COLLECTION OF POLLUTANTS

Often the most serious problem in air pollution control is collection of the pollutants so as to provide treatment. Automobiles are notorious polluters because their emissions cannot be readily collected. If we could channel the exhausts from automobiles to some central facilities, their treatment would be much more reasonable and feasible than controlling each individual car.

A successful example of collecting pollutants has been the recycling of blow-by gases in the internal combustion engine. By reigniting these gases and emitting them through the car's exhaust system, the necessity of installing a separate treatment device for the car has been eliminated.

Air pollution control engineers have their toughest problems when the pollutants from an industry are not collected but are emitted from windows, doors, and cracks in the walls, the so-called "fugitive emissions." It is not, therefore, always possible to solve a problem by simply installing some piece of control equipment. Often a complete overhaul of the air flow in the entire plant is required.

COOLING

The exhaust gases to be treated are sometimes too hot for the control equipment, and the gases must first be cooled. This can be done in three general ways: dilution, quenching, or heat exchange coils (Figure 20–2). Dilution is acceptable only if the total amount of hot exhaust is small. Quenching has the added advantage of scrubbing out some of these gases and particulates, but this method may result in a dirty and hot liquid that must be disposed of. The cooling coils are probably the most widely used and are especially appropriate when heat can be conserved.

TREATMENT

Selection of the correct treatment device requires matching characteristics of the pollutant with features of the control device. It is important to realize that air pollutants exhibit a wide range of physical properties and that the size of pollutants range many orders of magnitude. It is therefore not reasonable to expect one device to be effective and efficient for all pollutants. In addition, the types of chemicals in emissions often will dictate the use of some devices. For example, a gas containing a high concentration of SO_3 could be cleaned by water sprays, but the resulting sulfuric acid might present serious corrosion problems.

The various air pollution control devices are conveniently divided into those for controlling particulates and those used for controlling gaseous pollutants. The reason, of course, is the difference in the size. Gas molecules have diameters of about 0.0001 microns, particulates range from 0.1 microns up.

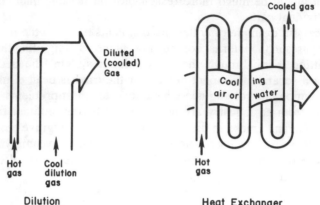

Dilution   Heat Exchanger

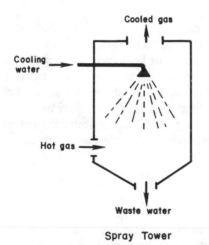

Spray Tower

Figure 20-2   Cooling hot waste gases.

Settling Chambers

The simplest devices for controlling particulates are settling chambers. These are nothing more than wide places in the exhaust flue where larger particles can settle out of the exhaust gas stream. They are usually equipped with a baffle to slow the gas stream. Obviously, only very large particulates ($>100\mu$) can be efficiently removed in settling chambers.

Cyclones

Possibly the most popular, economical, and effective means of controlling particulates is the cyclone. Figure 20–3 shows a simple schematic of a cyclone. The dirty air is blasted into a conical cylinder away from the centerline. This creates a violent swirl within the cone, and the heavy solids migrate to the wall of the cylinder by centrifugal action where they slow down due to friction and exit at the bottom of the cone. The clean air is in the middle of the cylinder and exits out the top. Cyclones are widely used as precleaners to remove large particles before further treatment.

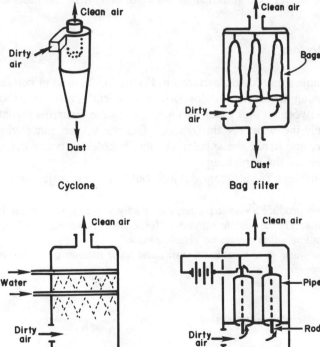
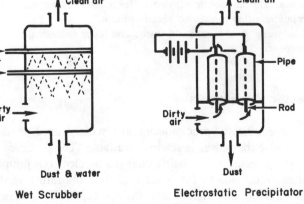

Cyclone   Bag filter

Wet Scrubber   Electrostatic Precipitator

Figure 20–3   Four methods of controlling particulates from stationary sources.

Bag (or Fabric) Filters

The filters used for controlling particulates (Figure 20–3) operate like the common vacuum cleaner. Fabric bags are used to collect the dust, which must be periodically shaken out of the bags. The fabric will remove nearly all particulates, including submicron-sized particles. Bag filters are widely used in many industrial applications, but are sensitive to high temperatures and humidity.

The basic mechanism of dust removal in fabric filters is thought to be similar to the action of sand filters in water quality management (see Chapter 5). The dust particles adhere to the fabric due to entrapment and surface forces. They are brought into contact by impingment and/or Brownian diffusion. Since fabric filters commonly have an air space-to-fiber ratio of 1:1, the removal mechanism cannot be simple sieving.

Wet Collectors

The simple spray tower, pictured in Figure 20–3, is an effective method for removing large particulates. More efficient scrubbers promote the contact between air and water by violent action in a narrow throat section into which the water is introduced. Generally, the more violent the encounter, and hence the smaller the gas bubbles or water droplets, the more effective the scrubbing.

Wet scrubbers are efficient devices but have two major drawbacks.

1. They produce a visible plume, albeit only water vapor that quickly disappears. The lay public seldom differentiates, however, and hence public relations often dictate no visible plume.
2. The waste is now in liquid form, and some manner of treatment is often necessary.

Electrostatic Precipitators

Today, electrostatic precipitators are widely used in power plants, mainly because the power is readily available. The particles are removed from the gas stream by first being charged by electrons jumping from one high-voltage electrode to the other and then migrating to the positively charged collecting electrode (electrons are negatively charged). The type of electrostatic precipitator shown in Figure 20–3 consists of a pipe with a wire hanging down the middle. The particulates collect on the pipe and must be removed by rapping the pipes with hammers. Electrostatic precip-

Figure 20–4   Effectiveness of electrostatic precipitators on a coal-fired power plant. The electrostatic precipitator on the right has been turned off.

itators have no moving parts, require only electricity to operate, and are extremely effective in removing submicron particulates. They are also expensive.

A major problem with electrostatic precipitators is that as the dust layer builds up inside the pipe, the resistivity increases, thus decreasing the drift velocity. Ironically, one means of reducing the insulating effect of the dust is to inject sulfur (!) as SO_2 or sulfuric acid into the gas.

Figure 20–4 shows the effectiveness of an electrostatic precipitator in controlling emission from a power plant. The large white boxes in the foreground are the electrostatic precipitators. The one on the right, leading to the two stacks on the right, has been turned off to show the effectiveness by comparison with the almost undetectable emission from the stacks on the left.

Comparison of Particulate Control Devices

The efficiencies and costs of the various control devices obviously vary widely. Figure 20–5 shows approximate collection efficiency curves, as a function of particle size, for the various devices discussed.

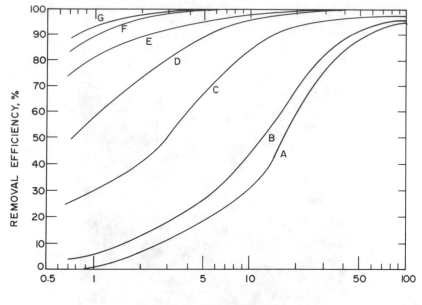

Figure 20-5   Comparison of removal efficiencies. A = baffled settling chamber, B = simple cyclone, C = high efficiency cyclone, D = electrostatic precipitator, E = spray tower wet scrubber, F = Venturi scrubber, G = bag filter.

CONTROL OF GASEOUS POLLUTANTS

The control of gases involves the removal of the pollutant from the exhaust gas stream, a chemical change in the pollutant, or a change in the process producing the pollutant.

Wet scrubbers, such as already discussed, can remove gaseous pollutants by simply dissolving them in the water. Alternatively, a chemical may be injected into the scrubber water, which then reacts with the pollutants. This is the basis for most SO_2 removal techniques as discussed below. *Adsorption* is a useful method when it is possible to bring the pollutant into contact with an efficient adsorber-like activated carbon. This method is effective for many organic pollutants (Figure 20-6.)

Incineration (Figure 20-7), or flaring, is used when an organic pollutant can be oxidized to CO_2 and water. A variation of incineration is *catalytic combustion* where the temperature of the reaction is lowered by the use of a catalyst that mediates the reaction.

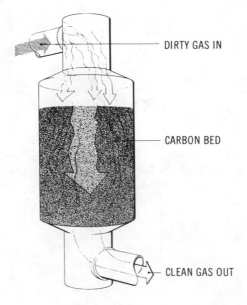

DIRTY GAS IN

CARBON BED

CLEAN GAS OUT

Figure 20-6 Adsorber for control of gaseous pollutants (courtesy American Lung Association).

Control of Sulfur Oxides

We note earlier that sulfur oxides (SO_2 and SO_3) are serious and yet ubiquitous air pollutants. The major source of sulfur oxides (or SO_x as they are often referred to in shorthand) is coal-fired power plants. The increasingly strict standards for SO_x control have prompted the development of a number of options and techniques for reducing emissions of sulfur oxides. Among these options are:

1. *Change to Low-Sulfur Fuel.* Natural gas is extremely low in sulfur. Oil burned for industrial heating and electric power production ranges in sulfur content from 0.5 to 3 percent, and coal varies from 0.3 to 4 percent. Low-sulfur fuels are an expensive and sometimes uncertain option.

2. *Coal Desulfurization.* Sulfur in coal can be both inorganically bound as iron pyrite, FeS_2, or organically bound. Pyrite can be removed by pulverizing the coal and washing it with a detergent solution. Organically bound sulfur can be removed by washing the coal with very concentrated acid. Preferred methods are coal gasification, which produces pipeline-quality gas, or solvent extraction, which produces low-sulfur liquid fuel.

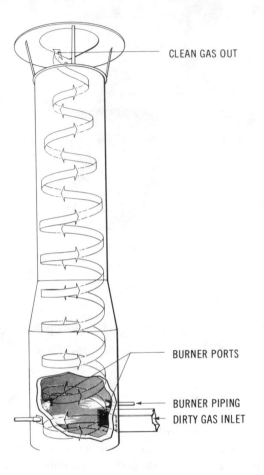

CLEAN GAS OUT

BURNER PORTS

BURNER PIPING
DIRTY GAS INLET

Figure 20–7   Incinerator for controlling gaseous pollutants (courtesy American Lung Association).

3. *Tall Stacks.* A short-sighted [pun intended] method, albeit locally economical, of SO_2 control is to build incredibly tall smokestacks and disperse the SO_x. This option has been employed in Great Britain and is in part responsible for the acid rain problem plaguing Scandinavia.

4. *Flue-Gas Desulfurization.* The last option is to reduce the SO_x emitted by cleaning the gases coming from the combustion, the so-called flue gases. Many systems have been employed, and a great deal of research is presently under way to make these processes more efficient.

During the past ten years, flue-gas desulfurization methods have been developed that trap SO_2 as the sulfite compound rather than a sulfate. The advantage of these methods is that the absorbing material can be regenerated so that the waste disposal problem is somewhat mitigated. A typical sulfite control method is single alkali scrubbing for which the reactions are

$$SO_2 + Na_2SO_3 + H_2O \rightarrow 2NaHSO_3$$

Sodium sulfite is then regenerated from the bisulfite by heating:

$$2NaHSO_3 \rightarrow Na_2SO_3(s) + H_2O + SO_2(conc)$$

The concentrated SO_2 recovered from this process can be used industrially (for example, in pulp and paper manufacture or for sulfuric acid manufacture).

Control of Nitrogen Oxides

Nitrogen oxides from fuel combustion are an important contributor to air pollution. Wet scrubbing, particularly with an alkaline scrubbing solution, will absorb NO_2. This method is, however, not a feasible one to install only for NO_2 control as in oil-burning power plants.

A moderately effective method, when scrubbing is not feasible, is off-stoichiometric combustion. This method controls NO formation by limiting the amount of air (or oxygen) in the combustion process to just a bit more than is needed to burn the hydrocarbon fuel in question. For example, the reaction for burning natural gas is

$$CH_4 + 2O_2 \rightarrow 2H_2O + CO_2$$

Nitrogen in the air used for combustion will compete with the natural gas for oxygen, in the competitive reaction

$$N_2 + O_2 \rightarrow 2NO$$

The stoichiometric ratio of oxygen needed in natural gas combustion is

$$32 \text{ g } O_2/16 \text{ g } CH_4$$

if the air used for combustion provides oxygen in slight excess of this ratio, virtually all the oxygen will combine with the fuel rather than with the nitrogen in the air. In practice, off-stoichiometric combustion is achieved by adjusting the air flow to the combustion chamber so that any persistent visible plume just disappears.

Control of Volatile Organics and Odors

Volatile organic compounds and odors are controlled by thorough oxidation either by incineration or by catalytic combustion.

CONTROL OF MOVING SOURCES

Although many of the above control techniques can apply to moving sources as well as stationary ones, one very special moving source—the automobile—deserves special attention. As noted in Chapter 18, the automobile has many potential sources of pollution. Realistically, however, there are only a few important points requiring control (Figure 20–8):

A.   evaporation of hydrocarbons (HC) from the fuel tank
B.   evaporation of HC from the carburetor
C.   emissions of unburned gasoline and partially oxidized HC from the crankcase
D.   the NO_x, HC, and CO from the exhaust

The most difficult control problem is the exhaust, which accounts for about 60 percent of the HC and almost all of the NO_x, CO, and lead.

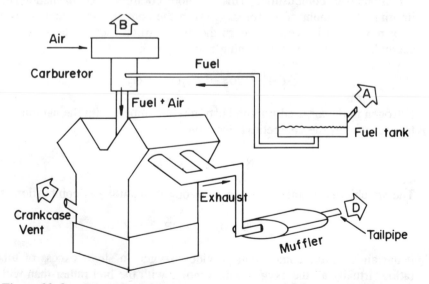

Figure 20–8   Internal combustion engine showing four major emission points.

Table 20–1.   Effect of Engine Operation on Automotive Exhaust
Characteristics, Shown as Fraction of Idling Emissions

	Component		
	CO	HC	NO_x
Idling	1.0	1.0	1.0
Accelerating	0.6	0.4	100
Cruising	0.6	0.3	66
Decelerating	0.6	11.4	1.0

One immediate problem is how to measure these emissions. It is not as simple as sticking a sampler up the tailpipe since the quantity of pollutants emitted changes with the mode of operation. The effect of operation emissions is illustrated as Table 20–1. Note that, when the car is accelerating, the combustion is efficient (low CO and HC), and the high compression produces a lot of NO_x. On the other hand, decelerating results in low NO_x and very high HC due to partially burned fuel.

Because of these difficulties the EPA has initiated a standard test for measuring emissions. This test procedure includes a cold start, acceleration, and cruising on a dynamometer to simulate a load on the wheels, and a hot start.

Emission control techniques for the internal combustion engine include tune-ups, exhaust gas recirculation, engine modifications, and catalytic reactors. A tune-up can have a significant effect on emissions. For example, a high air/fuel ratio (a lean mixture) will reduce both CO and hydrocarbons but will increase NO emission. A well-tuned car is nonetheless the first line of defense for emission control.

Both CO and hydrocarbons can be reduced by as much as 60 percent by recirculating the exhaust gas through the engine. There are two major drawbacks to this control method: the resulting control does not meet the 1983 U.S. standards and the recirculated gas must be cooled considerably before recycling or the piston surfaces will become deformed. Even with cooling there is unusual wear on the engine when this method is used, and it has largely been superseded.

The catalytic reactor ("catalytic converter") must perform two pollution-control functions: *oxidation* of CO and hydrocarbons to CO_2 and water and *reduction* of NO to N_2. Reduction of NO is accomplished by burning a fuel-rich mixture and thus depleting the oxygen to the catalyst: Introducing air and oxidizing the CO and HC at reduced temperatures suppresses the production of NO. In modern engines, a platinum-rhodium, three-stage catalyst is used, and both the air/fuel mixture and the temperature are carefully controlled.

An emission-free internal combustion engine is something of a contradiction in terms. Drastic lowering of emissions to produce a virtually pollution-free engine would require replacing the internal combustion engine with an external combustion engine. Laboratory tests indeed indicate that such engines can achieve better than 99 percent control of all three major exhaust pollutants. Although work began in 1968 on a mobile external combustion engine for modern cars, a working model has yet to be built. The unsolved problem is finding a working fuel (the "steam" in a "steam engine") that will permit ready acceleration but that is not flammable. Water, the working fuel in steam engines and in the old Stanley Steamer automobile, has a high heat capacity and heat of vaporization and thus does not respond quickly enough on acceleration. Organic fluids, which have the right combination of thermodynamic properties, are usually flammable.

Natural gas can be used as a fuel in cars, but there are so many competing uses for a limited supply that a changeover to natural gas as an automobile fuel would not be feasible. Electric cars are clean but can only store limited power; thus their range is limited. In addition, the electricity used to power such vehicles must be generated, thus creating more pollution.

The diesel engines used in trucks and buses also are important sources of pollution. There are, however, fewer of these vehicles in operation than there are gasoline-powered cars. The main problems associated with diesel engines are the visible smoke plume and odors, two characteristics that have led to considerable public irritation with diesel-powered vehicles. Diesel engines in passenger cars often meet emission control standards because diesel engines burn hotter and oxidize the fuel more efficiently.

CONCLUSION

This chapter is devoted mainly to the description of air pollution control alternatives by "bolt-on" devices. It should be reemphasized that this is usually the most expensive method of control. A general environmental engineering truism is that the least expensive and most effective control point is always the farthest up the process line. It is at the beginning of the process, or, better yet, consideration of alternatives to the process, where the most effective control is achieved. This is obvious in the case of flue-gas desulfurization. We should not seek to bury Rhode Island but rather ask if all the electricity is really necessary. Are there other power sources? Not only is this good engineering technology and economics, but it is also sensitive and enlightened analysis of our lifestyle and its impact on the environment.

PROBLEMS

20.1 Taking into account cost, ease of operation, and ultimate disposal of residuals, what type of control device would you suggest for the following emissions?
    a.   A dust with particle range of 5–10 μ?
    b.   A gas containing 20 percent SO_2 and 80 percent N_2?
    c.   A gas containing 90 percent HC and 10 percent O_2?

20.2 A stack emission has the following characteristics: 90 percent SO_2, 10 percent N_2, no particulates. What treatment device would you suggest and why?

20.3 How big is 10,000 microns in inches? How big is 1 micron in inches?

20.4 An industrial emission has the following characteristics: N_2—80 percent, O_2—15 percent, CO_2—5 percent. You are called in as a consultant to advise on the type of air pollution-control equipment required. What would be your recommendation?

20.5 Design a start/stop/drive test for measuring the emission of pollutants from an automobile. Justify your choice.

Chapter 21

Air Pollution Law

As with water pollution, a complex system of laws and regulations governs the use of air pollution abatement technologies. In this chapter, the evolution of air pollution law is described from its roots in common law through the passage of federal statutory and administrative initiatives. Problems encountered by regulatory agencies and polluters are addressed with particular emphasis on the impacts the system may or may not have on future economic development. Figure 21-1 offers a roadmap to be followed through this maze.

AIR QUALITY AND COMMON LAW

When dealing with common law, an individual or groups of individuals injured by a source of air pollution may cite general principles in two branches of that law that have developed over the years and may apply to their particular damages:

- tort law
- property law

The harmed party, the plaintiff, could enter a courtroom and seek remedies from the defendant for damaged personal well-being or damaged property.

Tort Law

A tort is an injury incurred by one or more individuals. Careless accidents, defamation of character, and exposure to harmful airborne chemi-

315

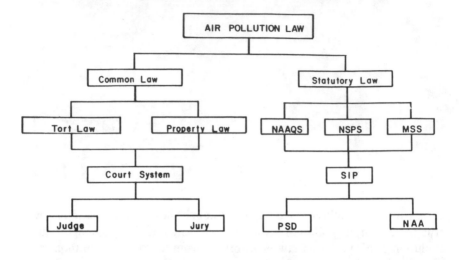

KEY:
NAAQS- National Ambient Air Quality Standards
NSPS – New (stationary) Source Performance Standards
MSS – Moving Source Standards
SIP   – State Implementation Plans
PSD – Prevention of Significant Deterioration
NAA – Non-Attainment Areas

Figure 21-1   Air pollution law in the United States.

cals are the types of wrongs included under this branch of common law. A polluter could be held responsible for the damage to human health under three broad categories of tort liability: intentional liability, negligence, and strict liability.

Intentional liability requires proof that somebody did a wrong to another party *on purpose*. This proof is especially complicated in the case of damages from air pollution. The fact that a "wrong" actually occurred must first be established, a process that may rely on direct statistical evidence or strong inference such as the results of lab tests on rats. In addition, intent to do the "wrong" must be established, which involves producing evidence in the form of written documents or direct testimony from the accused individual or group of individuals. Such evidence is not easily obtained. If intentional liability can be proven to the satisfaction of the courts, actual damages as well as punitive (punishment) damages can be awarded to the injured plaintiff.

Negligence may involve mere inattention by the air polluter who allowed the injury to occur. Proof in the courtroom focuses on the lack of reasonable care taken on the defendant's part. Examples of such neglect in air pollution include failure to inspect the operation and maintenance (O&M) of electrostatic precipitators or the failure to design and size an adequate abatement technology. Again, damages can be awarded to the plaintiff.

Property Law

Property law, on the other hand, focuses on the theories of nuisance and property rights; nuisance is based on the interference with the use or enjoyment of property, and property rights is based on actual invasion of the property. Property law is founded on ancient actions between land owners and involves such considerations as property damage and trespassing. A plaintiff basing a case on property law rolls the dice and hopes the court will rule favorably as it balances social utility against individual property rights.

Nuisance is the most widely used form of common law action concerning the environment. Public nuisance involves unreasonable interference with a right, such as the "right to clean air," common to the general public. A public official must bring the case to the courtroom and represent the public that is harmed by the air pollution. Private nuisance, on the other hand, is based on unreasonable interference with the use and enjoyment of private land. The key to a nuisance action is how the courts define "unreasonable" interference. Based on precedents and the arguments of the parties involved, the common law court balances the equities, hardships, and injuries in the particular case and rules in favor of either the plaintiff or the defendant.

Trespass is closely related to the theory of nuisance. The major difference is that some physical invasion, no matter how minor, is technically a trespass. Recall that nuisance theory demands an "unreasonable" interference with land and the outcome of a particular case depends on how a court defines "unreasonable." Trespass is relatively cut and dried. Examples of trespass include physical walk-ons, vibrations from nearby surface or subsurface strata, and possibly gases and microscopic particles flowing from an individual smoke stack.

In conclusion, common law has generally proven inadequate in dealing with problems of air pollution. The strict burdens of proof required in the courtroom often result in decisions that favor the defendant and lead to smoke stacks that continue to pollute the atmosphere. In addition, the technicality and complexity of individual cases often limit the ability of a court to act; complicated tests and hard-to-find experts often leave a court and a plaintiff with their hands tied. Furthermore, the absence of standing in a common law courtroom often prevents private individuals from bringing a case before the judge and jury unless the individual can demonstrate actual material or bodily harm from the air pollution.

One key aspect of these common law principles is their degree of variation. Each state has its own body of common law, and individuals relying on the court system are generally confined to using the common laws of the applicable state.

Given these shortcomings inherent in common law, Congress adopted a federal Clean Air Act, with the objective of plugging some of the pollution holes in common law and bringing consistency to air pollution control law.

STATUTORY LAW

Federal statutory law controlling air pollution began with the 1963 and 1967 Clean Air Acts. Although these laws provided broad clean air goals and research money, they did not apply air pollution controls throughout the entire United States, but only in heavily industrialized, dirty communities. In 1970, however, the Clean Air Act was amended to cover the entire United States, and the Environmental Protection Agency (EPA) was created to promulgate clean air regulations and to enforce the act. The 1970 amendments are the basis for the clean air legislation we have today. Additional significant amendments were made in 1977, and others are being considered in 1990.

National Ambient Air Quality Standards (NAAQS)

The EPA is empowered to determine allowable ambient concentrations of certain pollutants; these are the NAAQS. The primary NAAQS are intended to protect human health; the secondary NAAQS, to "protect welfare." The latter levels are actually determined as those needed to protect vegetation. These standards are listed in Table 21-1.

In 1988, EPA revised the standards for total suspended particulate matter. It has been observed that the particles that are most closely correlated with adverse health effects are those having a diameter of 10 μm or less. The revision of the NAAQS provides that:

- primary standards are set only for particles 10 μm or less in diameter. This is called the PM_{10} standard
- the primary annual average PM_{10} standard is 65 $\mu g/m^3$
- the primary 24-hour PM_{10} standard is 150 $\mu g/m^3$

Table 21-1.   Selected National Ambient Air Quality Standards (NAAQS)

Pollutant	Primary (ppm)	Primary (μg/m^3)	Secondary (ppm)	Secondary (μg/m^3)
Particulate Matter (μg/m^3)				
Annual geometric mean	—	75	—	60
Max 24-hr concentration	—	260	—	150
Sulfur Oxides				
Annual arithmetic mean	0.03	80	0.02	60
Max 24-hr concentration	0.14	365	0.1	260
Max 3-hr concentration	—		0.5	1,300
Carbon Monoxide				
Max 8-hr concentration	9	10,000	same as primary	
Max 1-hr concentration	35	40,000		
Photochemical Oxidants				
Max 1-hr concentration	0.12	210	same as primary	
Hydrocarbons				
Max 3-hr concentration	0.24	160	same as primary	
Nitrogen Oxides				
Annual arithmetic mean	0.05	100	same as primary	
Lead				
Avg of 3 months		1.5	same as primary	

It has been proposed that the secondary 24-hour standard be changed to a value between 70 and 90 μ/m^3.

NAAQS are set on the basis of extensive collections of information and data on the effects of these air pollutants on human health, ecosystems, vegetation, and materials. These documents are called "criteria documents" by the EPA, and the pollutants for which NAAQS exist are sometimes referred to as "criteria pollutants." Data indicate that all criteria pollutants have some threshold below which there is no damage. Recent epidemiological information has lowered this threshold for lead; further data might eliminate a threshold for lead entirely.

Under the Clean Air Act, most enforcement power is delegated to the states by EPA. The states, however, must show EPA that they can clean up the air to the levels of the NAAQS. This showing is made in each state's Air Quality Implementation Plan (AQIP), a document that contains all that state's regulations governing air pollution control including local regulations within the state. The AQIP must be approved by EPA, but once it is approved, it has the force of federal law.

The 1970 Clean Air Act Amendments envisioned that virtually all of the United States would meet the ambient standards by 1975. When it became evident that the 1975 deadline would not be met, Congress again amended the Act in 1977. Two new deadlines for meeting NAAQS were set: 1987 for mobile source-related criteria pollutants and 1985 for all other criteria pollutants. With the exception of about five communities in the United States like Los Angeles and Denver, these deadlines have been met.

Regulation of Emissions

Under the Clean Air Act, EPA can set *emission* standards only for new or modified stationary sources, not for existing sources that are already in operation. Sources of hazardous air pollutants are an exception. Emissions from existing stationary sources are regulated by the state through regulations in the AQIP.

Emission standards for new and modified stationary sources are called the new source performance standards (NSPS). They are determined *not* by the need to meet ambient standards but are set according to the best available emissions control technology, taking into account the cost of such control technology and its energy requirements. In sum, the NAAQS are health based while the NSPS are technology based. As new technology has become available, NSPS for various facilities have been revised. The most notable example of such a change was the revision of NSPS for coal-burning electric generating plants. The 1971 NSPS for SO_2 was 1.2 pounds SO_2 per million BTU heat input—a standard that could be met by using low sulfur coal. The 1977 NSPS is 70 percent reduction in SO_2 emissions, which requires flue-gas desulfurization.

The 1977 amendments to the Clean Air Act also prohibit the substitution of tall stacks or curtailment during bad dispersion conditions for actual emission controls. In 1984, 1988, and 1990 absolute limits on SO_2 emissions were proposed that required rollback of total SO_2 emissions in the heavily industrialized southern and midwestern United States. Such rollback requirements may soon become law.

Prevention of Significant Deterioration

In 1973, the Sierra Club sued EPA—and won—for failing to protect the cleanliness of the air in those parts of the United States where the air was cleaner than the NAAQS. In response, Congress included prevention of significant deterioration (PSD) in the 1977 Clean Air Act amendments.

Table 21-2.   Maximum Allowed Increases under PSD (Values are in $\mu g/m^3$)

| Class | Particulate Matter | | Sulfur Dioxide | | |
	Annual Mean	24-hr Max	Annual Mean	24-hr Max	3-hr Max
I	5	10	2	5	25
II	19	37	20	91	512
III	37	75	40	182	700

For PSD purposes, the United States is divided into Class I and II areas with the possibility of Class III designation for some areas. Class I includes the so-called "mandatory Class I" areas—all national wilderness areas larger than 5000 acres and all national parks and monuments larger than 6000 acres—and any area that a state or native American tribe wishes to designate Class I. The rest of the United States is Class II except that a state or tribe can petition EPA for redesignation of a Class II area to Class III.

To date, the only pollutants covered by PSD are sulfur dioxide and particulate matter. PSD limits the allowed increases in these as indicated in Table 21-2. In addition, visibility is protected in Class I areas.

An industry wishing to build a new facility must show, by dispersion modeling using a year's worth of weather data, that it will not exceed the allowed increment. On making such a showing, the industry receives a PSD permit from EPA. The PSD permitting system has had considerable impact on siting new facilities.

Hazardous Air Pollutants

The Clean Air Act recognizes that some substances emitted into the air are particularly hazardous, and the NAAQS for particulate matter is not restrictive enough. These hazardous air pollutants are listed under Section 112 of the act, and EPA is required to set ambient and emission standards at a level that "provides an ample margin of safety to protect the public health." These National Emission Standards for Hazardous Air Pollutants (NESHAPS) are determined for *all* sources—both new and existing—of the particular hazardous pollutant.

A hazardous air pollutant is one that might reasonably be anticipated to cause an increase in deaths or serious, incapacitating illness. To date, the following substances have been designated hazardous air pollutants:

arsenic
asbestos
beryllium
benzene
mercury
radioactive substances
vinyl chloride monomer

Several of these substances are carcinogens and as such are considered to have no threshold of effect. For these pollutants, an "ample margin of safety" cannot be assured, and risk associated with various degrees of control must be assessed. Congress is presently considering requiring control of about 300 toxic or hazardous air pollutants instead of requiring each one to be listed before control is required.

Mobile Sources

The federal government has preempted the setting of emission standards for mobile sources. Emission standards for gasoline-powered vehicles require a reduction in CO, hydrocarbons, and nitrogen oxides beginning with the 1979 model year and culminating in:

* 90 percent reduction in CO and hydrocarbon emission in the 1983 models
* 75 percent reduction in NO_x emissions in the 1985 models and an 87 percent reduction in the 1992 models.

Nonattainment

A region in which the NAAQS for one or more criteria pollutants are exceeded more than once or twice a year is called a "nonattainment area" for that pollutant. The Clean Air Act Amendments of 1977 require the following in a nonattainment area:

* If there is nonattainment of the lead, CO, or ozone standard, a traffic reduction plan and an inspection and maintenance program for exhaust emission control are required. Failure to comply results in the state's loss of federal highway construction funds.
* If there is nonattainment resulting from stationary source emission, an offset program must be initiated. Such a program requires that there be a rollback in emissions from existing stationary sources such that total emissions

• after the new source operates will be less than before. New sources in non-attainment areas must attain the lowest achievable emission rate (LAER) without necessarily considering the cost of such emission control.

CONCLUSION

Air pollution law is a complex web of common and statutory law. Although common law has offered and continues to offer checks and balances between polluters and economic development, shortcomings do exist. Federal statutory law has attempted to fill the voids and to a certain extent has been successful in cleaning the air. Engineers must be aware of the requirements placed on industry by this system of laws. Particular attention must be paid to the siting of new plants in different sections of the nation.

PROBLEMS

21.1 Acid rain is a mounting problem, particularly in the northeastern states. Discuss how this problem might be controlled under the system of common law. Compare this approach with the remedy under federal statutory law.

21.2 Pollution from tobacco smoke can significantly degrade the quality of certain air masses. Develop sample town ordinances to improve the quality of air: (1) inside city buses and (2) over the spectators at both indoor and outdoor sporting events. Rely on both structural and nonstructural alternatives; that is, solutions that mechanically, chemically, or electrically clean the air and solutions that prevent all or part of the pollution in the first place.

21.3 Would you favor an international law that permits open burning of hazardous waste on the high seas? Open burning refers to combustion without emission controls. Would you favor such a law if emissions were controlled? Discuss your answers in detail. Would you require permits that specify time of day of burning, distance from shore of burning, or banning certain wastes from incineration?

21.4 Emissions from a nuclear power generating facility pose a unique set of problems for local health officials. Such facilities typically have very low emissions, but each has the potential for a catastrophic discharge. What precautions would you take if you were charged with protecting the air quality of nearby residents? How would your set of rules consider: (1) direction of prevailing winds, (2) age of residents, and (3) siting new schools?

21.5 Assume you live in a small town that has two stationary sources of air pollution: a laundry/dry cleaning establishment and a regional hospital. Automobiles and front porches in the town have a habit of turning black literally overnight. What air laws apply in this situation and, given that they presumably are not being enforced, how would you: (1) determine if, in fact, ambient and/or emissions standards were being violated and, if either was, (2) advise the residents to respond to see that the laws are enforced?

LIST OF SYMBOLS

EPA	=	U.S. Environmental Protection Agency
LAER	=	lowest achievable emission rate
NAAQS	=	National Ambient Air Quality Standards
NESHAPS	=	National Emission Standards for Hazardous Air Pollutants
PSD	=	prevention of significant deterioration
AQIP	=	(State) Air Quality Implementation Plan

Chapter 22

Noise Pollution

The ability to make and detect sound provides humans with the facility to communicate with each other as well as to receive useful information from the environment. Sound can provide warning (the fire alarm), useful information (a whistling tea kettle), and enjoyment (music).

In addition to such useful and pleasurable sounds there is noise, often defined as unwanted or extraneous sound. What is and is not noise is often in dispute, especially in the area of music.[1]

We generally think of noise as an unwanted by-product of our civilization and classify such sources as trucks, airplanes, industrial machinery, air conditioners, and similar sound producers as noise.

Urban noise is not, surprisingly, a modern phenomenon. Legend has it, for example, that Julius Caesar forbade the driving of chariots on Rome cobblestone streets during nighttime so he could sleep.[2] Before the widespread use of automobiles, the 1890 noise pollution levels in London were described by an anonymous contributor to the *Scientific American* as:[3]

> The noise surged like a mighty heart-beat in the central districts of London's life. It was a thing beyond all imaginings. The streets of workaday London were uniformly paved in 'granite' sets . . . and the hammering of a multitude of iron-shod hairy heels, the deafening side-drum tattoo or tyred wheels jarring from the apex of one set (of cobblestones) to the next, like sticks dragging along a fence; the creaking and groaning and chirping and rattling of vehicles, light and

---

[1]George Bernard Shaw entered a posh London restaurant, took a seat, and was confronted by the waiter. "While you are eating sir, the orchestra will play anything you like. What would you like them to play?" Shaw's reply? "Dominoes."

[2]R.A. Baron, *The Tyranny of Noise* (New York: St. Martins Press, 1970).

[3]H. Still, *In Quest of Quiet* (Harrisburg, Pa.: Stackpole Books, 1970).

heavy, thus maltreated; the jangling of chain harness, augmented by the shrieking and bellowings called for from those of God's creatures who desired to impart information or proffer a request vocally— raised a din that is beyond conception. It was not any such paltry thing as noise.

Noise, as we will see, can adversely affect humans physiologically as well as psychologically. It is an insidious pollutant in that damage is usually long range and permanent. And yet it is certainly the pollutant of least public concern and (except for radioactivity) the least understood.

THE CONCEPT OF SOUND

The person-on-the-street has little, if any, concept of what sound is or what it can do. This is perhaps best exemplified by a well-meaning industrial plant manager who decided to decrease the noise level in his factory by placing microphones in the plant and channelling the noise through loudspeakers to the outside.[4]

Sound is in effect a transfer of energy. For example, rocks thrown at you would certainly get your attention, but this would require the transfer of mass (rocks). Alternatively, your attention can be gained by poking you with a stick, in which case the stick is not lost, but energy is transferred from the poker to the pokee. In the same way, sound travels through a medium such as air without a transfer of mass. Just as the stick had to move back and forth, so must air molecules oscillate in waves in order to transfer energy.

The small displacement of air molecules that creates pressure waves in the atmosphere is illustrated in Figure 22–1. As the piston is forced to the right in the tube, the air molecules next to it are reluctant to move and instead pile up on the face of the piston (Newton's First Law). These compressed molecules now act as a spring and release the pressure by jumping forward, creating a wave of compressed air molecules that move through the tube. The potential energy has been converted to kinetic energy.

These pressure waves move down the tube at a velocity of 344 m/sec (at 20 °C). If the piston oscillates at a frequency of say 10 cycles/sec, there will be a series of pressure waves in the tube each 34.4 m apart. This relationship is expressed as

$$\lambda = \frac{c}{f}$$

---

[4]R. Taylor, *Noise* (New York: Penguin Books, 1970).

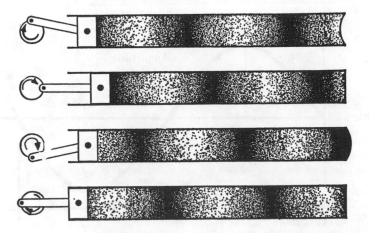

Figure 22-1   A piston creates pressure waves which are transmitted through air.

where   λ = wavelength, m
        c = velocity of the sound in a given medium, m/sec
        f = frequency, cycles/sec

Sound travels at different speeds in different materials, depending on the material's elasticity.

Example 22.1

In cast iron, sound waves travel at about 3440 m/sec. What would be the wavelength of a sound from a train if it rumbles at 50 cycles/sec and one listens to it placing an ear on the track?

$$\lambda = \frac{c}{f} = \frac{3440}{50} \cong 69 \text{ m}$$

In acoustics, the frequency as cycles per second is denoted by the name Hertz, and written Hz.[5] The common audible range for humans is between 20 and 20,000 Hz. The middle A on the piano, for example, is 440 Hz. The frequency is one of the two basic parameters that describe sound.

If the amplitude of a pressure wave of a pure sound (a sound with

---

[5]In honor of German Physicist Heinrich Hertz (1857–1894).

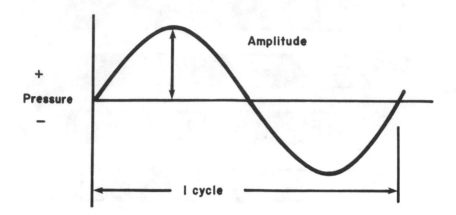

Figure 22-2   A representation of a sound wave.

only one frequency) is plotted against time, the wave is seen to produce a sinusoidal trace (Figure 22-2). All other nonrandom (dirty) sounds are made up of a number of suitable sinusoidal waves, as demonstrated originally by Fourier.

Although the human ear is a remarkable instrument, able to detect sound pressures over 7 orders of magnitude, it is not a perfect receptor of acoustic energy. In the measurement and control of noise, it is therefore important to know not only what a sound pressure is, but also to have some notion of how loud a sound *seems* to be. Before we address that topic, however, we must review some basics of sound.

SOUND PRESSURE LEVEL, FREQUENCY, AND PROPAGATION

Figure 22-2 represents a wave of pure sound: a single frequency. A sound wave is a compression wave, and the amplitude is a pressure amplitude measured in pressure units like N/m^2. As is the case with other wave phenomena, intensity is the square of the amplitude, or

$$I = P^2$$

The intensity of a sound wave is measured in watts.

When a person hears sounds of different intensities, the total intensity that is heard is not the sum of the intensities of the different sounds. Rather the human ear tends to become overloaded or saturated with too much sound. Another statement of this phenomenon is that human hearing sums up sound intensities logarithmically rather than linearly. A unit called the *bel*[6] was invented to measure sound intensity. Sound intensity level (IL) in bels is defined as

$$IL_b = \log_{10}(I/I_0)$$

where I = sound intensity in watts

I_0 = intensity or the least audible sound, usually given as 10^{-12} watts

The bel is an inconveniently large unit. The more convenient unit, which is now in common usage, is the *decibel (dB)*. Sound intensity level in dB is defined as

$$IL = 10 \log_{10}(I/I_0)$$

Since intensity is the square pressure, an analogous equation may be written for sound pressure level (SPL) in dB

$$SPL(\text{as dB}) = 20\log_{10} \frac{P}{P_{ref}}$$

where   SPL (dB) = sound pressure level in dB

P = pressure of sound wave

P_{ref} = some reference pressure, generally chosen as the threshold of hearing, 0.00002 N/m^2

These relationships are also derivable from a slightly different point of view. In 1825, E.H. Weber found that people can perceive differences in small weights, but if a person is already holding a substantial weight, that same increment is not detectable. The same idea is true with sound.

Although it is common to see the sound pressure measured over a full range of frequencies, it is sometimes necessary to describe a noise by the amount of sound pressure present at a specific range of frequencies. Such a *frequency analysis,* shown in Figure 22-3, can be used for solving industrial problems or evaluating the danger of a certain sound to the human ear.

In addition to amplitude and frequency, sound has two more characteristics of importance. Both can be visualized by imagining the ripples created by dropping a pebble into a large, still pond. The ripples, analogous to sound pressure waves, are reflected outward from the source, and the magnitude of the ripples is dissipated as they get farther away from the

---

[6]After Alexander Graham Bell.

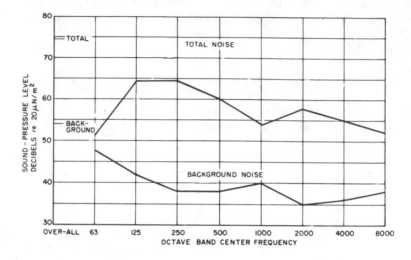

Figure 22–3   Typical frequency analysis of a machine noise with background noise.

source. Similarly, sound levels decrease as the distance between the receptor and source is increased.

In summary, then, the four important characteristics of sound waves are:

1. Sound pressure is the magnitude or amplitude of sound.
2. The pitch is determined by the frequency of the pressure fluctuations.
3. Sound waves propagate away from the source.
4. Sound pressure decreases with increasing distance from the source.

But these characteristics ignore the human ear. We know that the ear is an amazingly sensitive receptor, but is it equally sensitive at all frequencies? Can we hear low and high sounds equally well? The answer to these questions lead us to the concept of *sound level*, which is discussed in Chapter 23.

The mathematics of adding decibels is a bit complicated. The procedure can be simplified a great deal by using the graph shown as Figure 22–4. As a rule of thumb, adding two equal sounds increases the SPL by 3 db, and if one sound is more than 10 dB louder than a second sound, the contribution of the latter is negligible.

Background noise (or ambient noise) must also be subtracted from any measured noise. Using the above rule of thumb, if the sound level is more than 10 dB greater than the ambient level, the contribution can be ignored.

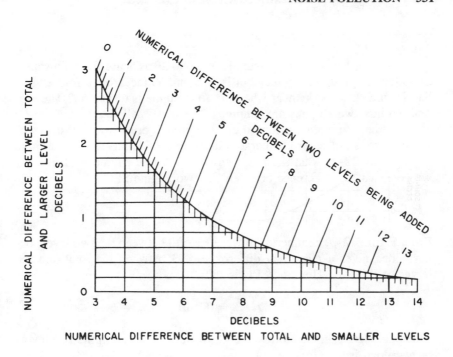

Figure 22–4 Chart for combining different sound pressure levels. For example: Combine 80 and 75 dB. The difference is 5 dB. The 5-dB line intersects the curved line at 1.2 dB, thus the total value is 81.2 dB. (Courtesy Gen Rad.)

The covering of sound with a louder one is known as masking. Speech can be masked by industrial noise, for example, as shown in Table 22–1. These data show that an 80-dB SPL in a factory will effectively prevent conversation. Telephone conversations are similarly affected with a 65-dB background making communication difficult and 80-dB making it impossible. In some cases, it has been found advisable to use *white noise,* a broad frequency hum, to mask other more annoying noises.

Table 22–1.   Sound Levels for Speech Making

Distance (ft)	Speech Interference Level (dB)	
	Normal	Shouting
3	60	78
6	54	72
12	48	66

Example 22.2

A jet engine has a sound pressure level of 80 dB as heard from a distance of 50 feet. A ground crew member is standing 50 feet from a four-engine jet. What SPL does she hear when the first engine is turned on? The second, so that two engines are running? The third? All four?

When the first engine is turned on, the ground crew hears 80 dB (provided there is not other comparable noise in the vicinity). To determine from Figure 22–3, what she hears when the second engine is turned on, we note that the difference between the two engine SPL is

$$80 - 80 = 0$$

From Figure 22–3, a numerical difference of 0 between the two levels being added gives a numerical difference of 3 between the total and the larger of the two. The total SPL is thus

$$80 + 3 = 83 \text{ dB}$$

When the third engine is turned on, the difference between the two levels being added is

$$83 - 80 = 3$$

yielding a difference from the total of 1.8, for a total SPL of

$$83 + 1.8 = 84.8 \text{ dB}$$

When all four engines are turned on, the difference between the sounds being added is

$$84.8 - 80 = 4.8$$

yielding a difference from the total of 1.2, for a total SPL of 86 dB.

THE ACOUSTIC ENVIRONMENT

The types of sounds around us vary from a Rachmaninoff concerto to the roar of a jet plane. Noise, the subject of this chapter, is generally considered to be unwanted sound or the sound incidental to our civilization that we would just as soon not have to put up with. The intensities of some typical environmental noises are shown in Figure 22–5.

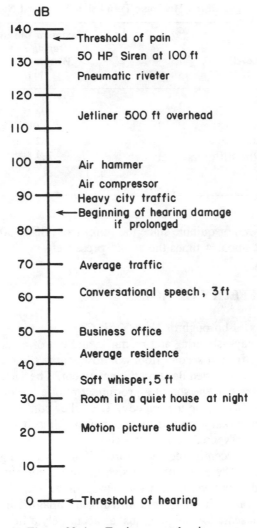

dB

140 — ←— Threshold of pain

130 — 50 HP Siren at 100 ft

Pneumatic riveter

120 —

110 — Jetliner 500 ft overhead

100 — Air hammer

Air compressor

90 — Heavy city traffic

←— Beginning of hearing damage
      if prolonged

80 —

70 — Average traffic

60 — Conversational speech, 3 ft

50 — Business office

Average residence

40 —

Soft whisper, 5 ft

30 — Room in a quiet house at night

20 — Motion picture studio

10 —

0 — ←—Threshold of hearing

Figure 22-4   Environmental noise.

In the industrial environment, noise is regulated by federal legislation called the Occupational Safety and Health Act (OSHA), which sets limits for noise in the workplace. Table 22–2 lists these limits.

There is some disagreement as to the level of noise that should be allowed for an 8-hour working day. Some researchers and health agencies insist that 85 dB(A)[7] should be the limit. This is not, as it might seem on

---

[7]The designation dB(A) is explained in Chapter 23.

Table 22–2.   OSHA Maximum Permissible Industrial Noise Levels

Sound Level, dB(A)	Maximum Duration During Any Working Day (hr)
90	8
92	6
95	4
100	2
105	1
110	½
115	¼

the surface, a minor quibble since the jump from 85 to 90 dB is actually an increase of about 4 times the sound pressure!

HEALTH EFFECTS OF NOISE

In the Bronx, a borough of New York City, one spring evening, four boys were at play—shouting and racing in and out of an apartment building. Suddenly, from a second-floor window, came the crack of a pistol. One of the boys sprawled dead on the pavement. The victim happened to be thirteen years old, the son of a prominent public leader, but there was no political implication in the tragedy. The killer confessed to police that he was a nightworker who had lost control of himself because the noise from the boys prevented him from sleeping.

We have only recently become aware of the devastating psychological effects of noise. The effect of excessive noise on our ability to hear, on the other hand, has been known for a long time.

The human ear is an incredible instrument. Imagine having to design and construct a scale for weighing just as accurately a flea or an elephant. Yet this is the range of performance to which we are accustomed from our ears.

A schematic of the human auditory system is shown in Figure 22–6. Sound pressure waves caused by vibrations set the eardrum (*tympanic membrane*) in motion. This activates the three bones in the middle ear—the *hammer, anvil,* and *stirrup*. These bones, when activated, physically amplify the motion received from the eardrum and transmit it to the inner ear. This fluid-filled cavity contains the *cochlea,* a snail-like structure in which the physical motion is transmitted to tiny hair cells. These hair cells deflect, much like seaweed swaying in the current, and certain cells are

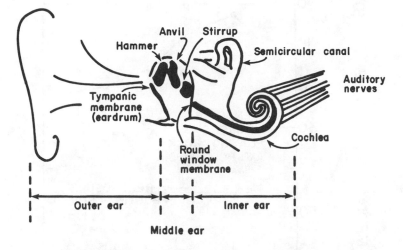

Figure 22-6   Cut-away drawing of the human ear.

responsive only to certain frequencies. The mechanical motion of these hair cells is transformed to bioelectrical signals and transmitted to the brain by the auditory nerves. Acute damage can occur to the eardrum, but this occurs only with very loud sudden noises. More serious is the chronic damage to the tiny hair cells in the inner ear. Prolonged exposure to noise of a certain frequency pattern can cause either temporary hearing loss, which disappears in a few hours or days, or permanent loss. The former is called *temporary threshold shift,* and the latter is known as *permanent threshold shift.* Literally, your threshold of hearing changes, so you are not able to hear some sounds.

Temporary threshold shift is generally not damaging to your ear unless it is prolonged. People who work in noisy environments commonly find that they hear less well at the end of the day. Performers in rock bands are subjected to very loud noises (substantially above the allowable OSHA levels) and commonly are victims of temporary threshold shift. In one study, the results of which are shown in Figure 22-7, the players suffered as much as 15 dB temporary threshold shift after a concert.

Repeated noise over a long time leads to permanent threshold shift. This is especially true in industrial applications where people are subjected to noises of a certain frequency. Figure 22-8 shows data from a study performed on workers at a textile mill. Note that the people who worked in the spinning and weaving parts of the mill, where noise levels are highest, suffered the most severe loss in hearing, especially at around 4000 Hz, the frequency of noise emitted by the machines.

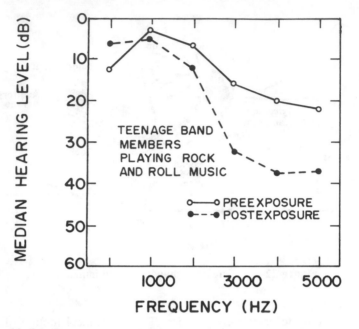

Figure 22–7   Temporary threshold shift for rock band performers. (Source: Data by the U.S. Public Health Service, in R. Taylor, *Noise* (New York: Penguin Books, 1970.)

CONCLUSION

Noise is a real and dangerous form of environmental pollution. Since people cannot adapt to it physiologically, we are perhaps adapting psychologically instead. Noise can keep our senses "on edge" and prevent us from relaxing. Our mental powers must therefore control this insult to our bodies. Since noise in the context of human evolution is a very recent development, we have not yet adapted to it and must thus be living on our buffer capacity. One wonders how plentiful this is.

PROBLEMS

22.1 If an office has a sound pressure level measured at 70 dB and a new machine emitting 68 dB is added to the din, what is the combined sound level?

22.2 The OSHA standard for 8-hr exposure to noise is 90 dB(A). The EPA suggests that this should be 85 dB(A). Show that the OSHA level is

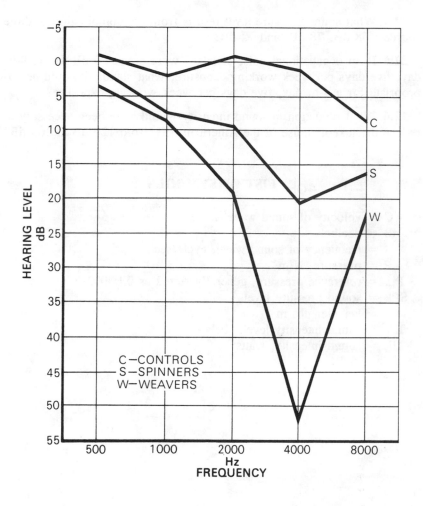

Figure 22–8 Permanent threshold shift in textile workers. [Source: Burns, W., *et al.* "An Exploratory Study of Hearing and Noise Exposure in Textile Workers," *Ann. Occup. Hyg.* 7, 323 (1958).]

almost 400 percent louder in terms of sound level than the EPA suggestion.

22.3 A machine with an overall noise level of 90 dB is placed in a room with another machine putting out 95 dB.

    a. What willl be the sound level in the room?

    b. Based on OSHA criteria how long should workers be in the room during one working day?

22.4 What sound pressure level results from the combination of three sources: 68 dB, 78 dB, and 72 dB?

22.5 If an occupational noise standard were set at 80 dB for an 8-hr day, five days per week working exposure, what standard would be appropriate for a 4-hr day, five days per week working exposure?

22.6 What reduction in sound intensity would have been necessary to reduce the takeoff noise of the American SST from 120 dB to 105 dB?

LIST OF SYMBOLS

C = velocity of sound wave, m/sec

dB = decibels

f = frequency of sound wave, cycle/sec

P = pressure, N/m^2

P_{ref} = reference pressure, generally stated as 0.00002 N/m^2

SPL = sound pressure level

λ = wavelength, m

IL_b = sound intensity levels, bels

IL = sound intensity, watts

Chapter 23

Noise Measurement and Control

Sound and noise are unique among environmental pollutants in that they can be measured both accurately and precisely. A number of techniques exist for measuring loudness, which in one way or another duplicate the logarithmic response of the human ear. These techniques may be used both to establish and to predict profiles of noise sources like airports.

SOUND LEVEL

Suppose you were put into a very quiet room and subjected to a pure tone at 1000 Hz at 40-dB SPL. If, in turn, this sound was turned off and a pure sound at 100 Hz was piped in and adjusted in loudness until you judged it to be "equally loud" to the 40-dB 1000-Hz tone you had just heard a moment ago, you would, surprisingly enough, judge the 100-Hz tone to be equally loud when it was at about 55 dB. In other words, more energy is needed at the lower frequency in order to hear a tone at the same loudness, indicating that the human ear is rather inefficient for low tones.

It is possible to conduct such experiments for many sounds and with many people, and to draw average equal loudness contours (Figure 23–1). These contours are in terms of *phons,* which are established relative to the sound pressure level in dB of the 1000-Hz reference tone.

Using Figure 23–1, a person subjected to a 65-dB SPL tone at 50 Hz would judge this to be equally as loud as a 40-dB 1000-Hz tone. Hence each tone has a loudness level of 40 phons.

Such measurements are not based only on basic physical phenomena, but also have a "fudge factor" thrown in that corresponds to the inefficiency of the human ear. Sound pressure levels that incorporate the inefficiencies of the human ear are called *sound levels.*

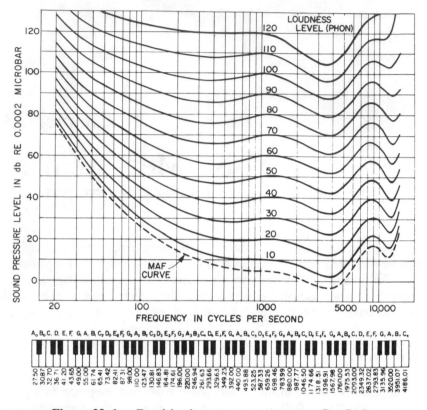

Figure 23–1   Equal loudness contours (courtesy Gen Rad).

Sound level (SL) is measured with a *sound level meter,* consisting of a microphone, amplifier, a frequency weighing circuit (filters) and an output scale. Figure 23–2 is a schematic diagram of a typical hand-held sound level meter. The weighing network filters out specific frequencies to make the response more characteristic of human hearing. Through use, three scales have become internationally standardized (Figure 23–3).

Note that the A scale in Figure 23–3 corresponds closely to an inverted 40-phon contour in Figure 23–1. Similarly, the B scale in reponse approximates an inverted 70-phon contour. The C scale is an essentially flat response, giving equal weight to all frequencies. It approximates the response of the ear to intense sound pressure levels.

The results of noise measurement using the standard sound level meter are expressed in terms of dB but with the scale designated. If on the A scale the meter reads 45 dB, the measurement is reported as 45 dB(A).

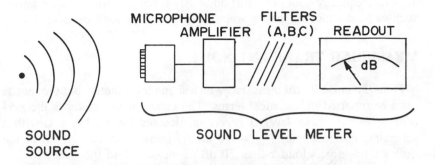

Figure 23-2   Schematic representation of a sound level meter.

Most noise ordinances and regulations are in terms of dB(A). This is a good approximation of human response for not very loud sounds. For very loud noise, the C scale is a better approximation, but because the use of multiple scales complicates matters, scales other than A are seldom used.

There is a further complication in measuring some noises, particularly those commonly called "community noise," such as traffic and loud parties. Although we have thus far treated noise as if it were constant in inten-

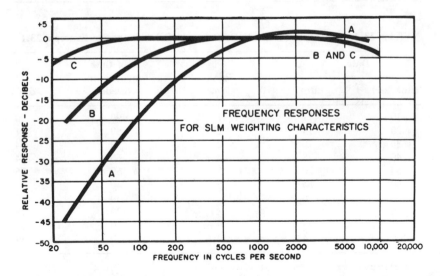

Figure 23-3   The A, B and C filtering curves for a sound level meter.

sity and frequency with time, this obviously is not true for transient noises such as trucks moving past a sound level meter or loud parties.

MEASURING TRANSIENT NOISE

Transient noise is still measured with a sound level meter, but the results must be reported in statistical terms. The common parameter is the *percent of time a sound level is exceeded,* denoted by the letter L with a subscript. For example, $L_{10} = 70$ dB(A) means that 10 percent of the time the noise was louder than 70 dB as measured on the A scale.

Transient noise data are gathered by reading the SL at regular intervals. These numbers are then ranked and plotted, and the L values read off the graph.

Example 23.1

Suppose the traffic noise data in Table 23–1 were gathered at ten-second intervals. These numbers are then ranked as indicated in the table and plotted as in Figure 23–5. Note that since ten readings were taken, the lowest reading (Rank No. 1) corresponds to a SL that is equalled or exceeded 90 percent of the time.

Hence, 70 dB(A) is plotted versus 90 percent in Figure 23–4.[1] Similarly, 71 dB(A) is exceeded 80 percent of the time.

Table 23–1.   Sample Traffic Noise Data and Calculations for Example 23.1

Data		Rank No.	% of time equal to or exceeded	dB(A)
Time (sec)	dBA)			
10	71	1	90	70
20	75	2	80	71
30	70	3	70	74
40	78	4	60	74
50	80	5	50	75
60	84	6	40	75
70	76	7	30	76
80	74	8	20	78
90	75	9	10	80
100	74	10	0	84

[1] You could also argue that since 70 dB(A) is the lowest value, it is the SL exceeded 100 percent of the time, and therefore, you could plot 70 dB(A) versus 100 percent. The second value, 71 dB(A), is exceeded 90 percent of the time, and so forth. Either method is correct, and the error diminishes as the number of data points increases. A third method is to plot $n/(m + 1)$ where n is the ranking and m the total number of data points (see Chapter 3).

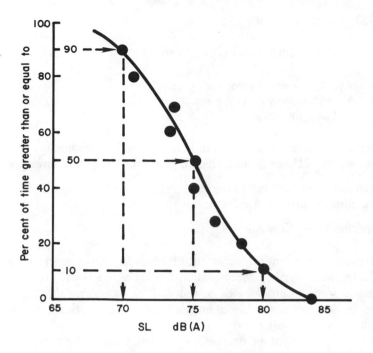

Figure 23-4   Results from a survey of transient noise (data from Table 23–1). The data are plotted as percent of time the sound level is exceeded, versus the sound level (SL) in dB(A).

One widely used parameter for gauging the perceived level of noise from transient sources is the noise pollution level (NPL), which takes into account the irritability of impulse (sudden) noises. The NPL is defined as

$$\text{NPL in dB(A)} = L_{50} + (L_{10} - L_{90}) + \frac{(L_{10} + L_{90})^2}{60}$$

As defined earlier, the symbol L refers to the percentage of time the noise is equal to or greater than some value. The percentage is indicated by the subscript. L_{10} is thus the dB(A) level exceeded 10 percent of the time.

With reference to Figure 23-4, L_{10}, L_{50}, and L_{90} are thus 80, 75, and 70 dB(A) respectively. These can be substituted into the equation and the NPL calculated.

It is always advisable to take as much data as possible. The ten readings illustrated above are seldom sufficient for a thorough analysis. The percentages must be calculated on the basis of the total number of readings taken. If, for example, 20 readings are recorded, the lowest (Rank No. 1) corresponds to 95 percent, the second lowest 90 percent, and so forth.

NOISE CONTROL

The control of noise is possible at three levels:

1.   reducing the sound produced
2.   interrupting the path of the sound
3.   protecting the recipient

When we consider noise control in industry, in the community, or in our homes we should keep in mind that all problems have these three possible solutions.

Industrial Noise Control

Industrial noise control generally involves the replacement of noise-producing machinery or equipment with quiet alternatives. For example, the noise from an air fan can be reduced by increasing the number of blades or the pitch of the blades and decreasing the rotational speed (thus obtaining the same air flow).

The second method of decreasing industrial noise is to interrupt the path, for example, by covering a noisy motor with insulating material.

The third method of noise control, often used in industry, is to protect the recipient by distributing earmuffs to the employees. But the problem is thus not really solved, and often the male workers refuse to wear earmuffs, considering them sissy. There seems to be something masculine about being able to "take it." Such misdirected masculinity will only end up in deafness.

The Occupational Safety and Health Act applies industrial noise limits to all jobs involving significant federal funds. These limits are tabulated in Chapter 22 as Table 22-2.

Community Noise Control

The three major sources of community noise are aircraft, highway traffic, and construction.

Construction noise must be controlled by local ordinances (unless federal funds are involved). Control usually involves the muffling of air compressors, jack hammers, hand compactors, and so forth. Since mufflers cost money, contractors will not take it upon themselves to control noise, and outside pressures must be exerted.

Aircraft noise in the United States is the province of the Federal Aviation Administration, which has instituted a two-pronged attack on this problem. First, it has set limits on aircraft engine noise and will not allow aircraft exceeding these limits to use the airports. This has forced manufacturers to design for quiet as well as for thrust.

The second effort has been to divert flight paths away from populated areas and to have pilots use less than maximum power when the takeoff carries them over a noise-sensitive area. Often this approach is not enough to prevent significant noise-induced damage or annoyance, and aircraft noise remains a real problem in urban areas.

The third major source of community noise is from highways. The car or truck creates noise by a number of means:

1. exhaust noise
2. tire noise
3. engine intake noise
4. gears and transmission
5. aerodynamic (wind) noise

A modern passenger car is so well muffled that its most important contribution, at moderate and high speeds, is tire noise. Sports cars and motorcycles, on the other hand, contribute exhaust, intake, and gear noise.

The worst offender on the highways is the heavy truck. Truck noise is generally from all of the above sources. In most cases, the total noise generated by vehicles can be correlated directly to the truck volume.

A number of alternatives are available for reducing highway noise. First, the source can be controlled by making quieter vehicles; second, highways could be routed away from populated areas; and third, noise could be baffled with walls or other types of barriers. Other methods include lowering speed limits, designing for nonstop operation, and reducing all highways to less than 8 percent grade.

Vegetation, surprisingly, makes a very poor noise screen. The most effective buffers have been to raise or lower the highway, or to build physical barriers beside the road and thus screen the noise. All of these have limitations. For example, noise will bounce off the walls and create little or no noise shadow. In addition, walls hinder highway ventilation, thus contributing to the buildup of dangerous air pollutants.

The Department of Transportation has established design noise levels for various land uses as shown in Table 23–2.

Table 23–2.   Design Noise Levels Set by the Federal Highway Administration

Land Category	Design Noise Level, L_{10}	Description of Land Use
A	60 dB(A) Exterior	Activities requiring special qualities of serenity and quiet, such as amphitheatres
B	70 dB(A) Exterior	Residences, motels, hospitals, schools, parks, libraries
	55 dB(A) Interior	Residences, motels, hospitals, schools, parks, libraries
C	75 dB(A) Exterior	Developed land not covered in categories A and B
D	no limit	Undeveloped land

Noise in the Home

Private dwellings are getting noisier because of internally produced noise as well as external community noise. The list of gadgets in a modern American home reads like a list of Halloween noisemakers. Some examples of domestic noise are listed in Table 23–3.

Otherwise similar products of different brands often will vary significantly in noise levels. When shopping for an appliance, it is just as important therefore to ask the clerk "How noisy is it?" as it is to ask him "How much does it cost?" And if he looks at you as if you had two heads, explain to him that he should know the dB(A) at the operator's position for all of his wares. He may actually bother to find out.

Table 23–3.   Some Domestic Noisemakers

Item	Sound Level (dBA)
Vacuum cleaner (10 ft)	75
Inside quiet car (50 mph)	65
Inside sports car (50 mph)	80
Flushing toilet	85
Garbage disposal (3 ft)	80
Window air conditioner (10 ft)	55
Ringing alarm clock (2 ft)	80
Lawn mower (operator's position)	105
Snowmobile (driver's position)	120
Rock band (10 ft)	115

CONCLUSION

As long as noise was considered just another annoyance in a polluted world, not much attention was given to it. We now have enough data to show that noise is a definite health hazard and should be numbered among our more serious pollutants. It is possible, using available technology, to lessen this form of pollution. The solution, however, costs money, and private enterprise cannot afford to give noise a great deal of consideration until it is forced to by either the government or the consuming public.

PROBLEMS

23.1 Given the following noise data, calculate the L_{10} and L_{50}.

Time (sec)	dB(A)	Time (sec)	dB(A)
10	70	60	65
20	50	70	60
30	65	80	55
40	60	90	70
50	55	100	50

23.2 In addition to the data listed in Example 23.1 and Table 23-1, the following SL measurements were taken:

Time (sec)	dB(A)	Time (sec)	dB(A)
110	80	160	95
120	82	170	98
130	78	180	82
140	87	190	88
150	92	200	75

Calculate the L_{50}, L_{10}, and NPL using all 20 data points.

23.3 Suppose your dormitory is 200 yards from a highway. What truck traffic volume would be "allowable" in order to stay within the Federal Highway Administration guidelines?

23.4 If the SL were 80 dB(C) and 60 dB(A), would you suspect that most of the noise was of high, medium, or low frequency? Why?

23.5 Many animals hear better than humans. Dogs, for example, can hear sounds at pressures close to $2 \times 10^{-6} N/m^2$. What is this in decibels?

23.6 If you sing at a level 10,000 times greater than the power of the faintest audible sound, at which dB level are you singing?

23.7 How many times more powerful is a 120-decibel sound than a 0 decibel sound?

23.8 On a graph of dB versus Hz (10 to 50,000), show a possible frequency analysis for: (a) a passing freight train, (b) a dog whistle, and (c) "white noise."

23.9 In your room measure and plot the sound level in dB(A) of an alarm clock versus distance. At what distance will it still wake you if it requires 70 dB(A) to get you up? Draw the same curve outside. What is the effect of your room on the sound level?

23.10 A noise is found to give the following responses on a sound level meter: 82 dB(A), 83 dB(B), and 84 dB(C). Is the noise of a high or low frequency?

23.11 A machine produces 80 dB(A) at 100 Hz (almost pure sound).

a. Would a person who has suffered a noise-induced threshold shift of 40 dB at that frequency be able to hear this sound? Explain.
b. What would this noise measure on the C scale of a sound level meter?

LIST OF SYMBOLS

$$
\begin{aligned}
\text{dB} &= \text{decibel} \\
\text{Hz} &= \text{Hertz, cycles/sec} \\
L_x &= \text{x percent of the time the stated sound level (L) was} \\
&\quad \text{exceeded, percent} \\
\text{NPL} &= \text{noise pollution level} \\
P &= \text{pressure, N/m}^2 \\
P_{ref} &= \text{reference pressure, N/m}^2 \\
\text{SLM} &= \text{sound level meter} \\
\text{SPL} &= \text{sound pressure level}
\end{aligned}
$$

Chapter 24

Environmental Impact

On January 1, 1970, the National Environmental Policy Act (NEPA) was signed into law by President Richard M. Nixon. NEPA declared and codified a national policy to encourage productive and enjoyable harmony between people and their environment. The law established the Council on Environmental Quality (CEQ), which monitors environmental effects of federal activities and assists the president in evaluating environmental problems and determining the best solutions to these problems.

The "sleeper" provision of NEPA, which has become the act's most far reaching and important, is the requirement that federal agencies evaluate and assess the consequences of any proposed action on the environment. Section 102(2)(C) reads, in part:

> The Congress authorizes and directs that, to the fullest extent possible: (1) the policies, regulations and public laws of the United States shall be interpreted and administered in accordance with the policies set forth in this chapter, and (2) all agencies of the Federal government shall include in every recommendation or report or proposals for legislation and other major Federal actions significantly affecting the quality of the human environment, a detailed statement by responsible officials on—

(i)   the environmental impact of the proposed action

(ii)   any adverse environmental effects which cannot be avoided should the proposal be implemented

(iii)   alternatives to the proposed action

(iv)   the relationship between local short-term uses of man's environment and the maintenance of long-term productivity, and

(v)   any irreversible and irretrievable commitments of resources which would be involved in the proposed action should it be implemented.

In other words, any federally funded or licensed project must be accompanied by an environmental impact statement (EIS). An EIS must assess the environmental impacts of a proposed project or action and must discuss alternatives to the proposed project or action, including a "no action" alternative. An EIS is generally issued in draft form for public comment, which is then incorporated into the final EIS. An EIS for a federal program (as distinct from a single federal action) is generally referred to as a "programmatic" EIS; a "generic" EIS for a class of programs can also be issued.

The requirement of the EIS has introduced environmental factors and concerns into the decision-making machinery. It has also assured public discussion of environmental factors. Unfortunately, the EIS process does not always work as envisioned. All too frequently the alterntives considered have been only those that seemed at the outset to be the most feasible and reasonable, and wholly different and innovative ways of perceiving the problem and devising solutions to it are overlooked.

In addition, simple opposition to a project or action can be registered within the EIS process only in the guise of finding inadequacies in the EIS. The engineer will recognize quickly that a determined critic can always point out gaps or inadequate coverage of issues in any EIS. The draft EIS for the trans-Alaska pipeline, the criticism of this EIS, and the final EIS responding to the criticism total almost 30 volumes. This sort of subversion of the intent of the environmental assessment process has decreased markedly in recent years as federal court decisions have shaped and clarified the EIS process.

PROCEDURE FOR CONDUCTING AN EIS

Ideally, an EIS must be thorough, interdisciplinary, and as quantitative as possible. The writing of an EIS involves three distinct phases: *inventory, assessment,* and *evaluation.* The first is a cataloging of environmentally susceptible areas, the second is the process of estimating the impact of the alternatives, and the last is the interpretation of these findings.

ENVIRONMENTAL INVENTORIES

The first step in evaluating the environmental impact of a project or project's alternatives is to inventory factors that may be affected by the proposed action. In this step no effort is made to assess the importance of a variable. Any number, and many kinds, of variables may be included, such as:

1. the "ologies": hydrology, geology, climatology, and archeology
2. environmental quality: land, surface, and subsurface water, air, and sound
3. plant and animal life
4. socioeconomic factors

This step involves counting, measuring, and describing existing conditions.

ENVIRONMENTAL ASSESSMENT

The process of calculating projected effects that a proposed action or construction project will have on environmental quality is called environmental assessment. It is necessary to develop a methodical, reproducible and reasonable method of evaluating both the effect of the proposed project and the effects of alternatives that may achieve the same ends but that may have different environmental impacts. A number of semiquantitative approaches have been used, among them the checklist, the interaction matrix, and the checklist with weighted rankings.

Checklists are listings of potential environmental impacts, both primary and secondary. Primary effects occur as a direct result of the proposed project such as the effect of a dam on aquatic life. Secondary effects occur as an indirect result of the action. For example, an interchange for a highway will not directly affect a land area, but indirectly it will draw such establishments as service stations and quick food stores thus changing land use patterns.

The checklist for a highway project could be divided into three phases: planning, construction, and operation. During planning, consideration is given to environmental effects of the highway route and the acquisition and condemnation of property. The construction phase checklist will include displacement of people, noise, soil erosion, water pollution, and energy use. Finally, the operation phase will list direct impacts due to noise, air pollution, water pollution due to runoff, energy use, and so forth and indirect impacts due to regional development, housing, lifestyle, and economic development.

The checklist technique thus simply lists all of the pertinent factors; then the magnitude and importance of the impacts are estimated. The estimation of impact is quantified by establishing an arbitrary scale, such as:

0 = no impact
1 = minimal impact
2 = small impact
3 = moderate impact

4 = significant impact
5 = severe impact

This scale can be used to estimate both the magnitude and the importance of a given item on the checklist. The numbers can then be combined and a quantitative measurement of the severity of environmental impact for any given alternative estimated.

Example 24.1

A landfill is to be placed in a floodplain of a river. Estimate the impact using the checklist technique.

First the items impacted are listed, then a judgment concerning both importance and magnitude of the impact is made. In this example, the items are only a sample of the impacts one would normally consider. The numbers in this example are then multiplied and the sum obtained. Thus:

Potential Impact	*Importance* × *Magnitude*
Groundwater contamination	5 × 5 = 25
Surface water contamination	4 × 3 = 12
Odor	1 × 1 = 1
Noise	1 × 2 = 2
Total	40

This total of 40 can then be compared to totals calculated for alternative courses of action.

In the checklist technique most variables must be subjectively judged. Further, it is difficult to predict future conditions such as land-use pattern changes or changes in lifestyle. Even with these drawbacks, however, this method is often used by engineers in governmental agencies and consulting firms—mainly due to its simplicity.

The *interaction matrix* technique is a two-dimensional listing of existing characteristics and conditions of the environment and detailed proposed actions that may impact the environment. This technique is illustrated in Example 24.2. For example, the characteristics of water might be defined as:

- surface
- ocean
- underground
- quantity
- temperature

- groundwater rechange
- snow, ice, and permafrost

Similar characteristics must also be defined for air, land, and other important considerations.

Opposite these listings in the matrix are lists of possible actions. In our example, one such action is labeled *resource extraction,* which could include the following actions:

- blasting and drilling
- surface extraction
- subsurface extraction
- well drilling
- dredging
- timbering
- commercial fishing and hunting

The interactions, as in the checklist technique, are measured in terms of magnitude and the importance. The magnitudes are represented by the extent of the interaction between the environmental characteristics and the proposed actions and typically can be measured. The importance of the interaction, on the other hand, is often a judgment call on the part of the engineer.

If an interaction is present, for example between underground water and well drilling, a diagonal line is placed in the block. Numbers can then be assigned to the interaction, with 1 being a small and 5 being a large magnitude or importance, and these placed in the blocks with magnitude above and importance below. Appropriate blocks are filled in, using a great deal of judgment and personal bias, and then are summed over a line, thus giving a numerical grade for either the proposed action or environmental characteristics.

Example 24.2

Lignite (brown) coal is to be surface mined in the Appalachian Mountains. Construct an interaction matrix for the water resources (environmental characteristics) versus resource extraction (proposed actions).

We see that the proposed action would have a significant effect on surface water quality, and that the surface excavation phase will have a large impact. The value of the technique is seen when the matrix is applied to alternative solutions and individual elements in the matrix, as well as row and column totals are reviewed.

Proposed Action

Environmental Characteristics	Blasting + drilling	Surface excavation	Subsurface excavation	Well drilling	Dredging	Timbering	Commercial fishing	Total
Surface water	3/2	5/5						8/7
Ocean water								
Underground water		3/3						3/3
Quantity								
Temperature		1/2						1/2
Recharge								
Snow, ice								
Total	3/2	9/10						

This trivial example cannot fully illustrate the advantage of the interaction technique. With large projects having many phases and diverse impacts, it is relatively easy to pick out especially damaging aspects of the project as well as the environmental characteristics that will be most severely affected.

The search for a comprehensive, systematic, interdisciplinary, and quantitative method for evaluating environmental impact has led to the *checklist-with-weighted-rankings* technique. The intent here is to use a checklist as before in order to ensure that all aspects of the environment are covered, as well as to give these items a numerical rating in common units.

The first step is to construct a list of items that could be impacted by the proposed alternative, grouping them into logical sets. One grouping might be:

- Ecology
  Species and populations

       Habitats and communities
       Ecosystems
- Aesthetics
       Land
       Air
       Water
       Biota
       Man-made objects

- Environmental pollution
       Water
       Air
       Land
       Noise

- Human interest
       Educational and scientific
       Cultural
       Mood/atmosphere
       Life patterns

Each title might have several specific topics under it, for example under Aesthetics, "Air" may list: (1) odor, (2) sound, and (3) visual as items in the checklist.

We must now assign ratings to these items, in common units. One procedure is to first estimate the ideal, or natural levels of environmental quality (without man-made pollution) and take a ratio of the expected condition to the ideal. For example, if the ideal dissolved oxygen in the stream is 9 mg/L, and the effect of the proposed action is to lower the dissolved oxygen to 3 mg/L, the ratio would be 0.33. This is sometimes called the *environmental quality index* (EQI). Another option to this would be to make the relationship nonlinear, as shown in Figure 24–1. Lowering the dissolved oxygen by a few mg/L will not affect the EQI nearly as much as lowering it, for example, below 4 mg/L since a dissolved oxygen below 4 mg/L definitely has a severe adverse effect on the fish population.

EQIs are calculated for all checklist items, and the values are tabulated. Next the weights are attached to the items, usually by distributing 1000 parameter importance units (PIU) among the items. The product of EQI and PIU, called the environmental impact unit (EIU), is thus the magnitude of the impact multiplied by the importance.

This method has several advantages. We can calculate the sum of EIUs and evaluate the "worth" of many alternatives, including the do-nothing alternative. We can also detect points of severe impact, where the EIU after the project may be much lower than before, indicating severe degradation in environmental quality. Its major advantage, however, is that it

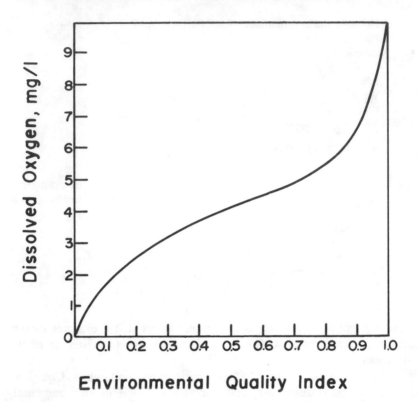

Figure 24–1   Projected environmental quality index curve for dissolved oxygen.

makes it possible to input data and evaluate the impact on a much less qualitative and a much more objective basis.

Evaluation

The final part of the environmental impact assessment is evaluation of the results of the preceding studies. Typically, the evaluation phase is out of the hands of the engineers and scientists responsible for the inventory and assessment phases. Decisions made within the responsible governmental agency ultimately use the EIS to justify past decisions or support new alternatives.

SOCIOECONOMIC IMPACT ASSESSMENT

Historically, the President's Council on Environmental Quality has been responsible for overseeing the preparation of EISs and CEQ regulations have listed what should be included in all EISs developed by federal agencies. In the case of the proposed projects discussed earlier in this chapter, two basic issues arose: public health dangers and environmental degradation. Under NEPA and the CEQ regulations, both issues must be addressed whenever alternatives are developed and compared.

In many cases, however, consideration of public health and environmental protection alone are not sufficient grounds on which to evaluate a range of alternative programs. Frequently, public acceptability is also a necessary input to an evaluation process. Although an alternative may protect public health and minimize environmental degradation, it may not be publicly acceptable, and thus could prove to be an essentially worthless alternative. Factors that influence public acceptability of a given alternative are generally discussed in terms of economics and broad social concerns. Economics includes the nationwide costs of an alternative, including the state, regional, local, and private components, the resulting impacts on user charges and prices, and the ability to finance capital expenditures. Social concerns include public preferences in siting (for example, no local landfills in wealthy neighborhoods) and public rejection of a particular disposal method (for example, food-chain landspreading of municipal sludge rejected on "general principle"). Consequently, each alternative developed to address the issues of public health and environmental protection must also be analyzed in the context of rigid economic analyses and broad social concerns.

Large changes in the population of a community, such as influxes of temporary construction workers or establishment of a military base with immigration of the associated personnel and their families, can have a number of impacts, both positive and adverse. New service jobs may well be created particularly in small communities, but there may also be increases in the crime rate, need for police and fire protection, and so forth. Study of such "boom town" phenomena has led to inclusion of these assessments in any EIS.

CONCLUSION

Engineers are required to develop, analyze, and compare a range of solutions to any given environmental pollution problem. This range of alternatives must be viewed in terms of their respective environmental impacts and economic assessments. A nagging question exists throughout any such viewing: can individuals really measure, in the strict "scientific" sense, degradation of the environment? Can we place a value on an unspoiled wilderness area? Unfortunately, qualitative judgments are required to assess many impacts of any project. This balancing of values is discussed in the final chapter.

PROBLEMS

24.1 Develop and apply an interaction matrix for the following proposed actions designed to clean municipal wastewater in your home town: (1) construct a large-scale activated sludge facility, (2) require septic tanks for households and small-scale package treatment plants for industries, (3) construct decentralized, small-scale treatment facilities across town, (4) adopt land application technology, (5) continue direct discharge into the river. Draw conclusions from the matrix.

24.2 Discuss the advantages and disadvantages of a benefit-cost ratio in deciding if a town should build a wastewater treatment facility. Focus on the valuation problems associated with analyzing the impacts of such a project.

24.3 Compare the environmental impact of a coal-fired electricity generating plant with those of a nuclear power plant. In your presentation, look at the flow of fuel from its natural state to the facility. Finalize your comparison with waste disposal considerations.

Chapter 25

Environmental Ethics

In Chapter 1 we point out that the control of environmental pollution is important since contamination of our environment can be dangerous to our health and increase the risk of mortality from environmental causes. We also say that this is only one of two major driving forces for controlling environmental pollution, the other force being a concern for the environment quite apart from the detrimental effect of contamination on our health, whether short- or long-term. In this chapter the concern with environmental contamination on our health is continued, followed by the introduction of the second reason for our environmental concern.

HUMAN BENEFIT FROM NOT POLLUTING THE ENVIRONMENT

Before the middle of the last century, people did not view the environment as being of any benefit to them. It was, in fact, a force to be fought with, to be subdued and conquered. Trees, rivers, and rocks were "the other," and human beings did not belong in this nature and were clearly superior to it. Nature was the enemy, and humans had a continual battle with its forces for survival.

By the mid 1800s, writers like Ralph Waldo Emerson began to argue that nature was not some evil force but rather a great provider and that we should be grateful for the many blessings nature yielded for our benefit. Emerson wrote even as the great rape of the environment was in progress, with forests and mineral and fossil fuel deposits being exploited without regard to the future. Stewart Udall called this the "myth of superabundance";[1] we could not *possibly* run out of any of these resources—

---

[1]Stewart Udall, "The Quiet Crisis" (Reading, Mass.: Addison-Wesley 1968).

there was just so *much* of them. In North Carolina, when a farmer tilled a piece of ground for three years and the tobacco stripped it of nutrients, he simply moved on to the next plot of ground. There was more than enough to go around for everyone.

Now we view such exploitation with horror and recognize that nature has its *instrumental value*. Nature should not be adversely treated because it is of value to us. Likewise we now recognize that global problems such as the depletion of the ozone layer and the destruction of forests in the Amazon basin can result in permanent harm to us or to our progeny.

Why exactly we care about future generations is an interesting philosophical question. Each generation inherits the successes and failures of the preceding generation. We naturally want our children's lives to be better than ours. But *why?* After all, what have they ever done for us?

Some philosophers suggest that, since we cannot know what the future will be like and what the needs of the future generations will be, that the only consideration we owe them is not to plan for them because we might mess it up. This argument is countered by the recognition that, even though we do not know in detail what these generations will be like, we can assume that they will still value *some* things we today recognize as good. Such concerns as health, freedom, equality, and so forth are values universal to the human experience, and there is no reason to believe that future generations will not also strive for such goods. Thus creating toxic disposal sites for future generations to clean up is *not* in their interest, regardless of how we might view these civilizations.

Nevertheless, our concern for future generations is a *human* concern, a concern for our own. It is not a concern for nature generally for its own sake. Interestingly, this latter concern is perhaps a far stronger and more potent force in the environmental movement than is our concern for our own health and safety. This second impetus for preserving our environment can best be described as an *environmental ethic*.

DUTIES TO NONHUMAN BEINGS: THE ENVIRONMENTAL ETHIC

The environmental ethic is a totally new concept, perhaps the most important new idea to surface in philosophy since the recognition that *all* people (including slaves) are deserving of our moral attention. The environmental ethic simply broadens the *moral community* to include not only people but *all of nature*. In other words, environmental ethics hold that we, people, have a moral responsibility to all animals, plants, and perhaps even inanimate objects even though they cannot return such moral responsibility. The argument that environmental ethics is nonsense because

nonpeople cannot reciprocate our concern is weak. Since we have no difficulty agreeing that little babies, comatose adults, and other humans who
cannot return our moral concern should be included in our moral community, why should whales, squirrels, or even pine trees not be included?

The environmental ethic is not site specific. No particular tree or even
forest is of more concern than others. Environmental ethics are not to be
confused with the "not-in-my-back-yard" syndrome. If a single tree is to
be included in the moral community, then *all* trees must be.

Such a concept is truly a revolutionary idea. Yet if we give it some
thought, it seems simply natural and meaningful. In this short chapter we
do not have space to develop all of the arguments for the existence of the
environmental ethic, but suffice it to say that the environmental ethic is
steadily becoming a part of our everyday lives. Concerns with endangered
species, for example, are not mainly because of their instrumental value
to humans but rather due to our sense that these species have a right to
exist. That is, they have an *intrinsic value* in addition to their possible instrumental value.

In a way the belief in the intrinsic value of nature represents a faith—one
has to feel that it is right. This feeling has been even translated to legalistic
terms most notably by C.D. Stone.[2] The question we face is whether
nature is here for the welfare of people, or whether it has an independent
claim to existence.

The problem is further complicated by what "nature" means. It is not
difficult to get people upset over the senseless slaughter of whales or baby
seals. Why? Simply because whales are too much like *humans;* they have
many of our most admirable characteristics and lead a placid, peaceful
life, and baby seals are so *cute!*

Can you imagine an equal outcry raised against the destruction of the
polio virus or the rattlesnake? Do these two species also have standing,
a right to be left alone and not be destroyed?

The issue is, of course, complex. But to not face it is unacceptable.
There is in the intrinsic approach to the value of nature a germ of a
positive and intuitively desirable notion. By not striving to develop and
foster this idea, might we be freezing our environmental ethic at too low
a level? Many widely accepted principles of the past, such as slavery and
the divine right of monarchs, have fallen. Albert Schweitzer noted that
Europeans long considered people with darker skins to be subhuman. It
was at one time considered stupid to think of them as equals and to treat
them humanely. Schweitzer expands this development to all of nature:

---

[2]C.D. Stone, *Should Trees Have Standing?—Toward Legal Rights for Natural Objects*
(Los Angeles, Calif.: Wm. Kaufman, Inc., 1972).

Today it is thought an exaggeration to state that a reasonable ethic demands constant consideration for all living things down to the lowest manifestations of life. The time is coming, however, when people will be amazed that it took so long for mankind to recognize that thoughtless injury to life was incompatible with ethics.[3]

Rene Dubos takes the idea further. Why, Dubos argues, do we limit our concerns with animate beings? He proposes, in a thoughtful article entitled "The Theology of the Earth" that everything has its place and reason for being. He suggests that there is a genius of the place, a oneness between people and uniqueness of each locality, be it a city or a grove of trees.[4]

The intrinsic value of nature obviously is on a higher ethical plane than is the instrumental view of nature, but it is a much more difficult idea for the mass of humanity to accept.

Yet if we look about us, we see evidence of mass destruction of our environment and a continued disregard of the earth's resources. How is it that many of our fellow human beings have never considered the concept of environmental ethics, much less heard the term? What are the roots of this behavior?

CAUSES OF THE PROBLEM OF OUR INCOMPATIBILITY WITH NATURE

Why is it that the human being—the most intelligent creature that ever evolved (or was created, depending on your beliefs)—is so incompatible with the environment?

There are three basic lines of argument in attempting to answer that question. One view is that humans created religions with dogmas that held the basic seeds of incompatibility. The second argument is that the social structure created by people makes it inherently impossible to attain equilibrium. The third view is that the growth of science and technology is responsible for environmental degradation.

It should be obvious that all three—religion, society, and technology—are intertwined and cannot be conveniently separated and individually scrutinized. Nevertheless, for the purpose of our discussion, let us assume that we can pick on each, one at a time.

---

[3]C. Joy, ed. "The Animal World of Albert Schweitzer," quoted in "The Intrinsic Value of Nature," D. Worster, *Env. Rev.* 4. no. 1 (1981).

[4]Rene Dubos, "A Theology of the Earth," a lecture reprinted in *Western Man and Environmental Ethics,* I.G. Barbour, ed., (Reading, Mass.: Addison-Wesley, 1973).

Religion as the Cause

Although the environment has been severely impacted by human cultures and civilizations, probably the greatest overall environmental damage has been caused in modern times. The major western religious traditions are rooted in Judaism and Christianity and these Judeo-Christian traditions have been blamed as the root cause of our environmental problems. And there is clearly some justification in this argument. In the first chapter of Genesis, for example, people are commanded by God to subdue nature, to procreate and have dominion over all living things. Such an anthropocentric view of nature runs all through the Judeo-Christian doctrine, a point forcefully made by Lynn White in 1967 in his essay "The Historical Roots of our Ecologic Crisis." [5]

White argues that the people embracing the Judeo-Christian religions are taught to treat nature as an enemy, that the religious dogmas prescribe that nature and natural resources are to be used only to meet the goals of survival and propagation. The religion of Western civilization, according to White, is responsible for environmental degradation. [6]

The Christian world reacted predictably to White's essay with books and papers claiming that Christianity is proenvironment and that the environmental ethic can in fact be found in the Bible. This is stretching it a bit, however.

The strongest argument presented to counter Lynn White's indictment of the Judeo-Christian doctrines as a root cause of our environmental problems centers on the notion of stewardship. Stewardship represents a view that people were put on the planet as caretakers, to see to the well-being of the earth. This "garden mentality" still relies on the concept of antropocentrism, however, a world of hierarchical relationships in which the human being is the noble, managing proprietor, accountable only to the sovereign. Such a doctrine also requires a substantial measure of faith in a transcendent God and a belief in a reward structure in the afterlife.

Perhaps the defensiveness of the Church toward Lynn White's ideas was unnecessary. White's assertion that religion molds moral is oversimplified, and blaming one religious tradition for our problems seems to be unfair. In fact, during the time that the Christian religion was becoming popular, people had many religious sects as alternatives, and the Christian ideas and ethics derived from ancient Judaic traditions seemed to fit most

---

[5] The article first appeared in *Science* 155 (10 March 1967): 1203. It has been reproduced in many books including *Western Man and Environmental Ethics,* I.G. Barbour, ed. (Reading, Mass.: Addision-Wesley, 1973).

[6] Or, as Mark Twain put, "Sometimes it seems a shame that Noah and his party did not miss the boat."

comfortably with the needs and existing value systems (whether active or latent). In short, the ethics existed first, and it was the people who developed these ethics into a religion.

Further, the ethics that fit the Judeo-Christian religions so well also spawned a vigorous pursuit of science, a tradition of democracy, and the capitalistic system. Although these religious principles were compatible with individual worth and achievement, the religions were not the reason that these ideas flourished. It can therefore be argued that the Judeo-Christian religions are not directly responsible for the traditions and ethics that promoted the destruction of nature.

Social Structure as the Cause

A second view of the roots of our ecological crisis is that our social structures are responsible. Probably the most damning piece yet written that takes this view is Garrett Hardin's "Tragedy of the Commons."[7]

Hardin illustrates his point by a story of a village that has a common green for the grazing of cattle surrounded by individual farmhouses. In the beginning, each farmer has one cow, and the green is able to support the herd. It becomes apparent to each farmer, however, that if he gets another cow, the cost is negligible to him personally (it is shared by everyone) but the profits are his alone. So he gets more cows, reaping greater and greater profits until the commons are no longer able to support the herd, and the system collapses.

Hardin used this parable to illustrate the problem of overpopulation, but it applies equally to other environmental problems. The social structure in the parable is of course capitalism, the individual ownership of wealth, and the use of that wealth for furthering one's own interests.

Much has been written about capitalism as the major causative agent for our environmental ills, often with the implication that some form of socialism, Marxist or otherwise, is a superior system. Unfortunately, the advantages of a centrally controlled (and hence totalitarian) system have not provided the answer. In fact, the environmental devastation in the U.S.S.R. is substantially more serious than it is in the West. When production is the primary goal of society, the environment and human life take a poor second place. Further, the recent attempt by China and the U.S.S.R. to modernize attests to not only the instability of their system but also to the existence of universal human aspirations and drives to better one's condition.

---

[7]*Science* 162 (1968):1243–1248, and reprinted in numerous books on environmental ethics.

The only types of socio-political systems that seemed at first glance to have developed a quasi-steady-state condition are primitive societies, such as the American Indians, the Finno-Ugric people of Northern Europe (Finland and Estonia) and the Maoris of New Zealand.

To all these people nature holds within it spirits that are both powerful and friendly (if at times capricious). The spirits in nature do not take human forms (as in Greek and Roman religions). Yet it is possible to converse freely with the spirits. The old Estonians and Finns, for example, always explained to the spirit of the tree why it was necessary to cut it down.[8] Such a reverence for, and closeness with, nature is unknown in most modern societies. Imagine the difficulty in the clearing of a forest for an artificial lake if every tree required a special explanation and apology?

There was, in these primitive societies, a camaraderie with all life forms. For example, old Estonians would begin the wheat harvest by cutting a shaft of wheat and placing this aside for the field mice. There did not seem to be any religious significance to this mouse-shaft (*hiirevihk*) and the only explanation handed down through the generations is that the mice deserve their share of the harvest.[9]

And yet these primitive societies were not all environmentally stable. The Maori first came to New Zealand in the 1300s and proceeded to exterminate the moa, a large ostrich-like bird that was their sole source of meat (the islands had no native mammals, except bats).

In the face of such failures, it is difficult to argue that we should reestablish primitive societies. In addition, a society is after all the reflection of the needs and aspirations of the people. People establish societies, and thus it cannot be argued that social systems are the root cause of our environmental ills.

Science and Technology as the Cause

It has become somewhat fashionable to blame our increased knowledge of nature and our related ability to put the knowledge to work for our environmental ills. The popularity of back-to-nature communes seems to suggest that somewhere we have made a wrong turn in the road to technologi-

---

[8]I. Paulson, *The Old Estonian Folk Religion* (Bloomington: Indiana University Press, 1971).
[9]Felix Oinas. Personal communication, Indiana University (1981).

cal capability. Numerous authors have jumped on this bandwagon and blamed technology for everything from athlete's foot to nuclear bombs. [10]

If technology is rightfully to blame, we must show first that other less technologically advanced societies avoided major environmental problems and secondly that modern technology is not value free.

The first premise is not true. Many primitive societies were equally, or even more, destructive than is our own. We already discussed the extermination of the moas by the Maori. Overgrazing in Africa and the destruction of forests by the early Greek civilization are other examples. In fact, one can mount a strong argument for technology as a means of being able to control destructive forces and trends.

Is technology value free? Is knowledge itself, without the application of it by humans, right or wrong, ethical or unethical? Before you jump to what might seem an obvious answer, consider this example: You have it within your power to conduct an experiment and thus prove the feasibility of constructing a simple and inexpensive nuclear bomb. The knowledge may well be used by terrorists in untold acts of violence. Without your experiment, the construction of such a device would not be possible. Should you conduct the experiment since you know that pure science and technology are value free and you will never put this knowledge to evil use? Or does this knowledge, by its very existence, constitute an evil?

Strong arguments have, as you can imagine, been advanced in defense of both views. But even if science is determined to be value laden, the use of science for curing environmental ills has an equally strong case. We cannot, in short, take the quick and dirty way out and blame science alone for environmental degradation just as we cannot lay the blame on religion or social structure alone.

THE FUTURE OF THE ENVIRONMENTAL ETHIC

The birth of the environmental ethic as a force is partly a result of our concern for our own long-term survival as well as our realization that humans are but one form of life and that we should share our earth with our fellow travelers.

One of the first to recognize the degradation of the environment and to

---

[10]The distrust of technological advancement is not new. During the Industrial Revolution in England, the Luddites were people who violently resisted the change from cottage industry to centralized factories. Because the large machines threatened their way of life, they smashed a few factories to make their point and were hanged for their trouble.

voice a concern for nature was Henry David Thoreau. His solution to this eloquently stated concern was withdrawal, which was perhaps morally admirable but realistically ineffective. What we have experienced in recent decades has been the coupling of Thoreau's concern with activism. It is one thing to be concerned, but it is much more effective to take action in order to promote your concern.

One powerful and original attempt at defining an environmental ethic was offered by Aldo Leopold, a naturalist and writer. He proposed that:

> A decision is right when it tends to preserve the integrity, stability, and beauty of the biotic community. It is wrong when it tends otherwise. [11]

Leopold argued convincingly toward the adoption of environmental values other than economic, and his writings have had a major impact on the growth of environmental awareness.

The environmental ethic is very new, and none of the doctrine is cast in immutable decrees and dogma. On the contrary, like all vital issues, the environmental ethic will undergo transformation as new data are made available and we humans are able to interpret more rationally and to live with nature.

Some critics have mistaken this maturation of the ethic for transience. The problem is so immense that to think of it as a fad is ludicrous (and fatal). But herein lies the dilemma: Unless there is public awareness as to the true nature of the problems and some realistic solutions, public concern may rapidly fade. For the impetus to survive, there must be public confidence, and the environmental scientist or engineer must engender this trust by analyzing and interpreting environmental problems correctly and proposing and developing constructed facilities compatible with our ecosystem. Sensationalism and the bandwagon tactics of some "environmentalists," politicians, and other well-meaning citizens can easily destroy the public sentiment so necessary for a successful assault on our common problems.

Education of the public to environmental problems and solutions (and nonsolutions) is of prime importance. It is necessary for people to be critical and literate in the scientific, technological, economical, and legal aspects of controlling environmental pollution. One must also recognize that an industrialized society does confer some benefits and that the "back-to-nature" movements of the late 1960s and early 1970s were not always environmentally beneficial. Often they failed to recognize carrying capacity and natural limits on population.

---

[11] A. Leopold, *A Sand County Almanac* (New York: Oxford University Press, 1949).

Recognizing the environmental ethic is not enough. We must also *live* the environmental ethic—recognize the power of nature and feel humble in the realization that we are just small cogs in a wonderful and still mysterious system.

Conversion Factors

Multiply	By	To Obtain
acre	43,560	ft^2
acre	0.404	ha
acre ft	1233	m^3
atmospheres	14.7	lb/in^2
British thermal units	252	cal
Btu	1.05×10^3	J
Btu/ft^3	8905	cal/m^3
Btu/lb	2.32	J/g
Btu/lb	0.555	cal/g
Btu/sec	1.05	kW
Btu/ton	278	cal/tonne
calories	4.18	joule
calories	3.9×10^{-3}	Btu
cal/g	1.80	Btu/lb
cal/m^3	1.12×10^{-4}	Btu/ft^3
cal/tonne (1000 kg)	3.60×10^{-3}	Btu/ton (2000 lb)
centimeters	0.393	in
feet	1.894×10^{-4}	mi
feet	0.305	m
ft/min	0.00508	m/sec
ft/sec	0.305	m/sec
ft^2	0.0929	m^2
ft^2	2.29×10^{-5}	acre
ft^3	0.0283	m^3
ft^3	28.3	liters

Multiply	By	To Obtain
ft^3	7.481	gal
ft^3/sec	0.0283	m^3/sec
ft^3/sec	449	gal/min
ft^3/sec	0.646	million gal/day
ft lb (force)	1.357	joule
ft lb (force)	1.357	newton meters
gallons	0.134	ft^3
gallons	3.78×10^{-3}	m^3
gallons	3.78	liters
gal/day/ft^2	0.0407	m^3/day/m^2
gal/min	2.23×10^{-3}	ft^3/sec
gal/min	0.0631	liter/sec
gal/min	1.44×10^{-3}	million gal/day
gal/min	0.227	m^3/hr
gal/min	6.31×10^{-5}	m^3/sec
gal/min/ft^2	2.42	m^3/hr/m^2
million gal/day	694	gal/min
million gal/day	43.8	liters/sec
million gal/day	3785	m^3/day
million gal/day	0.0438	m^3/sec
million gal/day	1.55	ft^3/sec
grams	2.2×10^{-3}	lb
hectares	2.47	acre
horsepower	0.745	kW
inches	2.54	cm
inches of mercury	0.49	lb/in^2
inches of mercury	3.38×10^3	newton/m^2
inches of water	249	newton/m^2
joule	0.239	calorie
joule	9.48×10^{-4}	Btu
joule	0.738	ft lb
joule	2.78×10^{-7}	kWh
joule	1	newton meter
J/g	0.430	Btu/lb
J/sec	1	watt
kilograms	2.2	lb (mass)
kg	1.1×10^{-3}	tons
kg/ha	0.893	lb/acre
kg/hr	2.2	lb/hr
kg/m^3	0.0624	lb/ft^3
kg/m^3	1.68	lb/yd^3
kilometers	0.622	mi

Multiply	By	To Obtain
km/hr	0.622	mph
kilowatts	1.341	horsepower
kWh	3600	kilojoule
liters	0.0353	ft^3
liters	0.264	gal
liters/sec	15.8	gal/min
liters/sec	0.0228	mgd
meters	3.28	ft
meters	1.094	yd
m/sec	3.28	ft/sec
m/sec	196.8	ft/min
m^2	10.74	ft^2
m^2	1.196	yd^2
m^3	35.3	ft^3
m^3	264	gal
m^3	1.31	yd^3
m^3/day	264	gal/day
m^3/hr	4.4	gpm
m^3/hr	6.38×10^{-3}	mgd
m^3/sec	35.31	ft^3/sec
m^3/sec	15,850	gpm
m^3/sec	22.8	mgd
miles	1.61	km
miles	5280	ft
mi^2	2.59	km^2
mph	0.447	m/sec
milligrams/liter	0.001	kg/m^3
million gallons	3785	m^3
mgd	43.8	liter/sec
mgd	157	m^3/hr
mgd	0.0438	m^3/sec
newton	0.225	lb (force)
newton/m^2	2.94×10^{-4}	inches of mercury
newton/m^2	1.4×10^{-4}	lb/in^2
newton meters	1	joule
newton sec/m^2	10	poise
pounds (force)	4.45	newton
pounds (force)/in^2	6895	N/m^2
pounds (mass)	454	g
pounds (mass)	0.454	kg
pounds (mass)/ft^2/yr	4.89	kg/m^2/yr
pounds (mass)/yr/ft^3	16.0	$kg/yr/m^3$
pounds/acre	1.12	kg/ha

Multiply	By	To Obtain
pounds/ft^3	16.04	kg/mg^3
pounds/in^2	0.068	atmospheres
pounds/in^2	2.04	inches of mercury
pounds/in^2	7140	newton/m^2
pounds/in^2	2.31	ft of water
tons (2000 lb)	0.907	tonne (1000 kg)
tons	907	kg
ton/acre	2.24	tonnes/ha
tonne (1000 kg)	1.10	ton (2000 lb)
tonne/ha	0.446	tons/acre
yd	0.914	m
yd^3	0.765	m^3
watt	1	J/sec

1 gallon water = 8.34 lb
1 ft^3 water = 62.43 lb
1 m^3 water = 2283 lb

Glossary and Abbreviations

Absorption: process by which one material is captured in another either chemically or by going into solution.

Activated Carbon: material made from coal by driving off hydrocarbons under intense heat but no oxygen, leaving a tremendous surface area on which many chemicals can adsorb.

Activated Sludge: suspension of microorganisms taken from the bottom of the final clarifier.

Activated Sludge System: consists of an aerated basin in which microorganisms are reducing organics to CO_2, H_2O, other stable materials, and more microorganisms; followed by a settling tank (final clarifier) in which the microorganisms are separated out and recirculated into the aeration basin.

Acute: severe and short-lived (a disease, for example).

Adiabatic Lapse Rate: refers to the change in temperature with elevation as the result of atmospheric pressure. In dry air the adiabatic lapse rate is $1\,°C/100$ m. Adiabatic means that heat is neither added nor removed.

Adsorption: process by which one material is attached to another such as an organic on activated carbon. It is a surface phenomenon.

Advanced Waste Treatment: wastewater treatment beyond the secondary or biological stage. It may include the removal of nutrients such as phosphorus and nitrogen or any other potential problems.

Aeration: the process of being supplied or impregnated with air. Aeration is used in wastewater treatment to foster biological purification.

Aerobic: presence of free oxygen.

Aerosol: suspension of fine particles in a gas.

Algae: one-celled aquatic organisms with chlorophyll that grow in the presence of light, CO_2, and nutrients and release oxygen.

Algal Bloom: a proliferation of living algae on the surface of lakes or ponds.

Alum: aluminum sulfate.

Alveoli: air sacs in the lung where oxygen and carbon dioxide transfer takes place.

Ambient Air: any unconfined portion of the atmosphere; the outside air.

Anaerobic: absence of free oxygen.

Anticyclone: high-pressure cell with winds circulating about a center (clockwise in the northern hemisphere).

Appropriations Doctrine: basis for water law in the western United States.

Aquifer: water-bearing geologic stratrum.

Asbestos: a mineral fiber with countless industrial uses; a hazardous air pollutant when inhaled.

Assimilation: the ability of a body of water to purify itself of organic pollution.

Asthma: difficulty in breathing caused by constriction of bronchial tubes.

Attrition: wearing or grinding down by friction. One of the three basic contributing processes of air pollution, the others are vaporization and combustion.

Audiometer: an instrument for measuring hearing sensitivity.

BOD: biochemical oxygen demand.

Bag Filter: device for removing particulates in an air stream.

Baling: a means of reducing the volume of solid waste by compaction.

Bar: unit of atmospheric pressure measurement equal to one dyne/cm^2. 1 bar = 1000 millibars.

Bar Screen: in wastewater treatment, a screen that removes large floating and suspended solids.

Benthic Region: the bottom of a body of water.

Beryllium: a metal that when airborne has adverse effects on human health; it has been declared a hazardous air pollutant.

Biodegradation: metabolic process by which high-energy organics are converted to low energy, CO_2, and H_2O.

Biochemical Oxygen Demand: amount of oxygen used by microorganisms (and by chemical reactions) in the biodegradation process. BOD is usually measured at 20 °C for 5 days.

Biodegradable: having the capacity of decomposing quickly as a result of the action of microorganisms.

Brackish Water: a mixture of fresh and salt water.

Bronchiole: small air tube in the lung.

Bronchitis: acute inflammation of air passages.

Bubbler: device for measuring gaseous air pollutants.

COD: chemical oxygen demand.

cfs: cubic feet per second; a measure of the amount of water passing a given point.

Calorie: the amount of heat required to raise the temperature of one gram of water one degree centrigrade.

Carcinogens: cancer-producing substances.

Catalyst: substance that speeds up a chemical reaction (such as combustion) without entering into the reaction.

Centrifuge: device for dewatering slurries such as wastewater sludge.

Chemical Oxygen Demand: amount of oxygen used in chemically oxidizing a substance.

Chlorinated Hydrocarbons: a class of generally long-lasting, broad-spectrum insecticides of which the best known is DDT.

Chlorination: the application of chlorine to drinking water, sewage, or industrial waste for disinfection or oxidation of undesirable compounds.

Chlorophyll: substance found in plants that absorbs energy from light.

Chlorosis: loss of green color.

Chronic: less severe and longer lived than acute.

Cilia: tiny hair cells lining air passages.

Clarifier: a settling tank.

Clear Well: storage tank for finished potable water.

Coagulation: process of chemically treating a turbid waste to reduce the charge on the particles and thus make it possible for them to flocculate.

Coliforms: group of bacteria that produce gas and ferment lactose some of which are found in the intestinal tracts of warm-blooded animals.

Combined Sewer: carries sanitary wastes as well as stormwater.

Comminutor: grinds up large solids as a preparation for further wastewater treatment.

Convection: transmission of energy by movement of fluids.

Cyclone: low-pressure system circulating winds (counterclockwise in the northern hemisphere).

Cyclone: device for removing large (5-micron) dust particles.

dB: decibel.

dB(A): decibels as measured on the A scale.

DDT: the first of the modern chlorinated hydrocarbon insecticides whose chemical name is 1,1,1-tricholoro-2,2-bis (p-chloriphenyl)-ethane.

DO: dissolved oxygen.

DOC: dissolved organic carbon.

Decibel: measure of sound intensity.

Decomposition: reduction of the net energy level and change in chemical composition of organic matter because of the actions of aerobic or anaerobic microorganisms.

Desalinization: salt removal from sea or brackish water.

Detention Time: average time required to flow through a basin.

Diffused Air: method of aerating the microorganisms in the aeration tank by blowing in air through porous diffusers.

Digestion: decomposition of organics, either aerobically or anaerobically, usually at high solids concentrations and, in the case of anaerobic digestion, at elevated temperatures.

Dust: fine solid particles.

Dustfall: gravimetric measurement of dust by settling in a jar.

Dystrophic Lakes: lakes between eutrophic and swamp stages of aging.

Ecology: the interrelationships of living things to one another and to their environment or the study of such interrelationships.

Edema: swelling of tissues and accumulation of fluid in a body or organ.

Effluent: liquid flowing out.

Electrostatic Precipitator: device for removing fine particulate matter in an air stream.

Emphysema: loss of air exchange capacity by deterioration of alveoli.

Epidemiology: science of statistically evaluating and dealing with diseases in populations.

Epilimnion: top layer of a lake.

Eutrophication: process of aging of lakes and other still water bodies; characterized by excessive aquatic growth.

Eutrophic Lakes: shallow lakes, weed choked at the edges and very rich in nutrients.

Fecal Coliforms: coliforms specifically originating from warm-blooded intestines.

Feedlots: concentrations of animals for fattening before slaughter.

Final Clarifier: last settling tank in wastewater treatment prior to discharge.

Flocculation: process of forming large clumps from small particles once the particles have coagulated.

Fluorosis: bone disease caused by excessive ingestion of fluorides.

Fly Ash: fine particles generated by noncombustible materials during the burning of coal.

Fume: dust forming from condensation of vapors.

Fungi: small plants without chlorophyll.

Garbage: food waste in refuse.

Genetic: pertains to origin and development.

Greenhouse Effect: warming due to trapping of heat radiation.

Grit Chamber: used to remove sand and other large, heavy particles in a wastewater treatment plant.

HC: hydrocarbons.

Hz: Hertz.

Heavy Metals: metallic elements with high molecular weights, generally toxic to plant and animal life in low concentrations. Examples include mercury, chromium, cadmium, arsenic, and lead.

Hemoglobin: iron-containing protein pigment in blood that carries oxygen from the lung to other parts of the body.

Heat Island: concentration of warm air over a city, preventing external circulation.

Hertz: a unit of frequency, in cycles/second.

Hi Vol: high volume sampler.

High-Volume Sampler: device used for measuring particulates in air by capture on a filter.

Humus: decomposed organic material.

Hydrocarbons: chemicals containing hydrogen and carbon.

Hypolimnion: bottom layer of a lake.

Incineration: oxidation in the presence of heat and free oxygen, ideally producing CO_2, H_2O, and other stable compounds.

Infiltration: water entering sanitary sewers through broken pipes, illegal connections, and so forth.

Influent: the liquid flowing in.

Interceptor Sewer: carries wastes from a sewerage system to a treatment plant or to another interceptor.

Inversion: atmospheric condition where a warmer body of air is above a colder air mass.

L_{10}: symbol for indicating 10 percent of data are greater than the stated number.

Lagoon: hole-in-the-ground for temporary disposal of wastes.

Lapse Rate: rate at which temperature varies with altitude.

Lime: calcium hydroxide.

Limnology: the study of the physical, chemical, meteorological, and biological aspects of fresh waters, specifically lakes.

MGD: millions of gallons per day, commonly used to express rate of flow.

Masking: covering over of one sound or odor by another.

Mechanical Aeration: method of aerating the microorganisms in the aeration tank by beating and splashing the surface.

Mercaptans: organic compounds containing sulfur.

Micrograms: one millionth of a gram.

Microscreening: removal of small particles from water by filtering through a metal screen on a rotating drum.

Mist: small liquid droplets in air.

Morbidity: measure of disease in a population.

Mortality: measure of death in a population.

NPL: noise pollution level.

Necrosis: death of living tissue.

Noise Pollution Level: calculated value in dB which takes into account the irritation of noise variation.

OSHA: Occupational Safety and Health Act.

Oligotrophic Lakes: deep lakes that have a low supply of nutrients and contain little organic matter.

Organophosphates: a group of pesticide chemicals containing phosphorus, such as malathion and parathion, intended to control insects.

Oxidation Pond: method of wastewater treatment allowing biodegradation to take place in a shallow pond.

Oxygen Sag: drop in DO following pollution of a stream with subsequent recovery.

ppm: parts per million; weight/weight for water, and volume/volume for air.

PAN: peroxyacetyl nitrate, a component of photochemical smog.

Particulates: finely divided solid or liquid particles in the air or in an emission.

Pathogen: microorganism that causes disease.

Percolation: movement of water from the surface into the ground.

Photochemical: pertaining to reactions affected by light.

Polyelectrolytes: chemicals used in water and wastewater treatment for coagulation and flocculation.

Potable Water: safe, drinkable, and pleasing water.

Primary Clarifier: first settling tank in a wastewater treatment plant.

Primary Treatment: removal of solids in wastewater.

Pulmonary: pertaining to lungs.

Pyrolysis: combustion in the absence of oxygen.

Rapid Mix: device used for mixing chemicals in water treatment.

Rapid Sand Filter: device for removing turbidity in water by seepage through a bed of sand and washing the sand by flow reversal.

Rainout: removal of air pollution by condensation into rain droplets.

Raw Sewage: untreated domestic or commercial wastewater.

Raw Primary Sludge: slurry from the bottom of the primary clarifier.

Refuse: urban solid waste.

Ringelmann: number used to report density of smoke.

Riparian Doctrine: base for water law in the eastern United States.

Rubbish: nongarbage fraction of refuse.

SL: sound level.

SPL: sound pressure level.

Salinity: the degree of salt in water.

Sanitary Landfill: solid waste disposal in the ground using approved techniques.

Sanitary Sewers: carry only domestic wastewater.

Scrubber: device for removing air contaminants by bringing them into contact with water.

Secondary Treatment: removal of oxygen demand in wastewater.

Sedimentation: settling of solids in water and wastewater.

Septic Tank: underground tanks for treating small flows of domestic wastewater.

Settling Tank: a tank in water and wastewater treatment where solids are allowed to settle.

Sewage: domestic wastewater.

Sewers: pipes used to carry wastewater.

Sewerage System: system of sewers used to carry wastewater.

Sludge: wastewater solids suspended in water.

Smog: originally a combination of smoke and fog, as occurred in London, now smog is synonymous with polluted air.

Smoke: waste from incomplete combustion expelled with the air stream.

Sound Level Meter: device for measuring sound in dB, either on A, B, or C scales.

Sound Level: sound approximately perceived by human ears, expressed as a reading on the Sound Level Meter.

Sound Pressure: fluctuating air pressure as propagated sound waves.

Sound Pressure Level: change in pressure due to a sound, measured in dB.

Spray: large droplets of liquid.

Storm Sewers: carry stormwater only.

Synergism: the cooperative action of separate substances so that the total effect is greater than the sum of the effects of the substances acting independently.

Tertiary Treatment: follows the secondary part of wastewater treatment and is used to polish the effluent.

Thermocline: inflection point in a lake temperature profile.

Threshold: limit below which no effect is discernible.

Tile Field: pipes laid in ground with spaces in between so as to promote percolation of wastewater into the ground.

Tone: pure sound uniform in frequency.

Trachea: air duct from larynx to bronchial tubes.

Transpiration: process of water transport to the atmosphere through plants.

Trickling Filter: device for removing oxygen demand from wastewater by dribbling the water over rocks covered with a zoological slime.

Turbidity: interference with the passage of light through water caused by suspended matter.

Turnover: mixing of a lake due to thermal variations.

USPHS: United States Public Health Service.

USPHSDWS: USPHS Drinking Water Standards.

Vacuum Filter: device used for dewatering wastewater sludge.
Venturi Scrubber: high-efficiency device for scrubbing contaminated air.
WHO: World Health Organization.
Washout: removal of air pollutants by rain.
Weir: metal plate over which liquid effluent flows; used in clarifiers.
Wind Rose: graphic representation of wind data.
Zooplankton: tiny aquatic animals.

Index

Acid formers, 123
Acid rain, 57, 252–253
Acoustic environment, 332–334
Acrolein, 260
Activated carbon adsorption, 112
Activated sludge system, 105
Aeration tank, 105, 106–107
Aerobic decomposition, 22–23
Aerobic stabilization, 122, 123
Agricultural wastes, 17
Agriculture, and nonpoint source
   pollution, 137–138
Air classifiers, 182
Air pollution, 241–265
  atmospheric dispersion of, 277–279
  atmospheric effects of, 263
  clean air definition and, 243
  cleansing atmosphere of, 279–281
  control of, 299–313
  early examples of, 2, 241–242
  hazardous waste and, 193, 196–197,
   231, 233, 254
  health effects of, 254–262
  horizontal dispersion of, 28
  laws governing, 315–323
  materials effected by, 263–264
  parts of problem with, 267
  sludge disposal and, 130
  source correction of, 299–300
  sources of, 249–254
  treatment of, 301–305
  types of, 243–249
  vertical dispersion of, 269–277
  water pollution and, 141
Air quality
  laws covering, 315–323
  management of, 268
  measurement of, 289–296

  solid waste disposal and, 233
Air Quality Implementation Plan
  (AQIP), 319, 321
Aldehyde, 260
Algal blooms, 32–34
Alpha (α) radiation, 203, 204–205, 211
Alum (aluminum sulfate), 80, 81
Ambient air quality, 11, 289
Ambient water quality stream
  classifications, 150
Ammonia
  aerobic decomposition and, 23
  anaerobic decomposition and, 23–24
  water quality measurement with,
   59–61
Anaerobic decomposition, 22, 23–25
  lakes and, 31–34
  landfills and, 171, 173
  streams and, 25–30
Anaerobic stabilization, 122, 123–125
Anticyclones, 268
Appropriations doctrine, 147–149
Aquatic ecology, 18–21
Aquifer, 69–70
Arsenic, 34, 63, 155, 263, 322
Asbestos, 254, 322
Ash, 176, 193, 194, 199
Assessment, environmental, 351–357
Atmosphere
  air pollutants dispersion in, 263,
   269–279, 285–287
  cleansing, 279–281
  water in, 276–277
Atomic adsorption spectrophotometry,
  63
Automobiles
  emissions from, 244–245, 250, 254,
   300–301, 310

water pollution and, 16, 86, 152
Infiltration, 86
Inflow, 86
Information, clearinghouses, hazardous
    waste, 191
Intensity level (IL), 329
Interceptors, 87
Inventories, environmental, 350–352
Inversion, 272, 273
Ionizing chambers, 211
Ionizing radiation, see Radiation
Irrigation, 114–115

Jackson Candle Turbidimeter, 55

Kjeldahl nitrogen, 59

Lakes
    acid rain and, 252–253
    water pollution and, 30–34
Land
    noise pollution and, 345–346
    riparian doctrine and, 146
Land cover, and water pollution, 136
Land disposal
    radioactive waste, 221
    sludge, 131–133
Land farms, 188
Landfills
    hazardous waste and, 197–200, 231
    leachate from, 172, 198–199
    solid waste and, 165, 170–171
    stages of, 172, 173
Land treatment of wastewater, 114–115
Land use planning, 141–142, 192
Latency period, 9
Laws
    air pollution, 315–323
    bottle recycling, 166
    environmental impact, 349–358
    hazardous waste, 185, 233–238
    nonhazardous solid waste, 232–233
    radioactive waste, 221–223
    water pollution, 145–156
    see also specific laws
LC_{50}, 11
LD_{50}, 11
Leachate, 172, 198–199
Lead, 34, 244, 254, 263, 264,
    310–311
Linear energy transfer (LET), 209, 211

Liquid scintillation, 218
Litter, 141, 165–166
Loading, 106
Local undesirable land uses (LULU), 8
Low-level waste (LLW), 215
Lungs
    air pollution and, 255
    particulate matter and, 258–260, 291
    radioactive waste and, 220

Magnets, 182–183
Management facilities
    hazardous waste, 192–200
    radioactive waste, 220–223
Manholes, 86, 87
Manifest, hazardous waste, 188–189
Masking, 330–331
Mass burn units, 183
Medicine, and radioactive waste, 218,
    223
Mercury, 34, 63, 254, 322
Mesophilic digestion, 126
Meteorology, 267–268, 289
Methane, 24, 172, 173
Methane formers, 123–124
Methyl isocyanate, 255
Methyl mercury, 34
Mining, 18, 140–141, 215
Mixed liquor suspended solids (MLSS),
    106
Mixed waste, 193
Modular incinerators, 174
Molybdenum, 263
Mortality, and risk analysis, 4–5, 8
Municipal solid waste (MSW),
    159–166
    characteristics of, 160–161
    collection of, 161–164
    disposal of, 165, 169–176, 231
    health problems and, 159–160
    impacts of, 232–233
    laws governing, 232–233
    quantities of, 160
    water pollution and, 16

National Ambient Air Quality
    Standards (NAAQS), 318–320, 322
National Emission Standards for
    Hazardous Air Pollutants
    (NESHAPS), 321